Jumpstart! Stu

Jumpstart! Study Skills presents a collection of engaging and simple-to-use games and activities that will jumpstart students' understanding of working independently and as part of a group and how to develop the study skills necessary to succeed at school.

Providing teachers with a range of practical ideas and strategies to promote active learning in Key Stage 2, the activities in this book will help students to:

- create plans for investigations and assignments;
- improve their organisational skills: time management and teamwork;
- collect data using methods such as observations, surveys and interviews;
- develop their reading and notetaking skills;
- engage in meaningful discussions and develop their talk skills;
- advance their computer skills to sift and record data;
- create strategies for revising and preparing for tests;
- analyse data and draw conclusions;
- improve their ability to write reports;
- evaluate their own achievements and identify future targets.

Jumpstart! Study Skills is an essential classroom resource that will encourage children's development and help teachers to deliver effective lessons that promote active learning in Key Stage 2.

John Foster taught English for twenty years before becoming a full-time writer and educational consultant. He has written over 100 books for classroom use and is a highly regarded children's poet and educational adviser.

Jumpstart!

Jumpstart! Wellbeing
Games and activities for ages 7–14
Steve Bowkett and Kevin Hogston

Jumpstart! Apps
Creative learning, Games and
activities for ages 7–11
*Natalia Kucirkova, Jon Audain and Liz
Chamberlain*

Jumpstart! Grammar (2ⁿᵈ Edition)
Games and activities for ages 6–14
Pie Corbett and Julia Strong

Jumpstart! Talk for Learning
Games and activities for ages 7–12
John Foster and Lyn Dawes

Jumpstart! PSHE
Games and activities for ages 7–13
John Foster

Jumpstart! History
Engaging activities for ages 7–12
*Sarah Whitehouse and Karen
Vickers-Hulse*

Jumpstart! Geography
Engaging activities for ages 7–12
Sarah Whitehouse and Mark Jones

**Jumpstart! Thinking Skills and
Problem Solving**
Games and activities for ages 7–14
Steve Bowkett

Jumpstart! Maths (2nd Edition)
Maths activities and games for ages
5–14
John Taylor

Jumpstart! Spanish and Italian
Engaging activities for ages 7–12
Catherine Watts and Hilary Phillips

Jumpstart! French and German
Engaging activities for ages 7–12
Catherine Watts and Hilary Phillips

Jumpstart! Drama
Games and activities for ages 5–11
*Teresa Cremin, Roger McDonald, Emma
Goff and Louise Blakemore*

Jumpstart! Science
Games and activities for ages 5–11
Rosemary Feasey

Jumpstart! Storymaking
Games and activities for ages 7–12
Pie Corbett

Jumpstart! Poetry
Games and activities for ages 7–12
Pie Corbett

Jumpstart! Creativity
Games and activities for ages 7–14
Steve Bowkett

Jumpstart! ICT
ICT activities and games for ages
7–14
John Taylor

Jumpstart! Numeracy
Maths activities and games for ages
5–14
John Taylor

Jumpstart! Literacy
Key Stage 2/3 literacy games
Pie Corbett

JUMPSTART!

STUDY SKILLS

GAMES AND ACTIVITIES FOR ACTIVE LEARNING, AGES 7–12

John Foster

Routledge
Taylor & Francis Group

LONDON AND NEW YORK

First published 2017
by Routledge
2 Park Square, Milton Park, Abingdon, Oxon OX14 4RN

and by Routledge
711 Third Avenue, New York, NY 10017

Routledge is an imprint of the Taylor & Francis Group, an informa business

British Library Cataloguing in Publication Data
A catalogue record for this book is available from the British Library

Library of Congress Cataloging in Publication Data
Names: Foster, John, 1941 October 12– author.
Title: Jumpstart! study skills : games and activities for active learning, ages 7–12 / John Foster.
Other titles: Study skills
Description: Abingdon, Oxon; New York, NY: Routledge, 2017.
Identifiers: LCCN 2016052618| ISBN 9781138241473 (hardback) | ISBN 9781138241480 (pbk.) | ISBN 9781315280219 (ebook)
Subjects: LCSH: Study skills—Study and teaching—Activity programs. | Active learning. | Activity programs in education.
Classification: LCC LB1601 .F67 2017 | DDC 371.30281--dc23
LC record available at https://lccn.loc.gov/2016052618

ISBN: 978-1-138-24147-3 (hbk)
ISBN: 978-1-138-24148-0 (pbk)
ISBN: 978-1-315-28021-9 (ebk)

Typeset in Palatino and Scala Sans
by HWA Text and Data Management, London

Printed and bound by CPI Group (UK) Ltd, Croydon, CR0 4YY

Contents

Introduction

Active learning has for a long time been an approach adopted in early years teaching. This book is designed to help children aged 7 to 11 to develop their study skills and thus to promote active learning in Key Stage 2.

Jumpstart! Study Skills provides opportunities for children to be fully involved in their learning by presenting them with problems to solve. Children learn by doing, thinking, exploring and sharing ideas. As they progress through primary school, active learning assists them in developing the study skills they require for their secondary schooling.

To succeed in their schoolwork, children need to develop the ability to plan how to carry out an investigation and to suggest how they might solve a problem. They need to be able to clarify what a task involves, to generate ideas and to discuss how they might find information, approach a task or solve a problem. They need to be able to ask relevant questions, to predict outcomes and consider consequences.

Having formulated a plan, children must be able to locate and collect relevant data, carry out research on the internet and gather information from books, using techniques such as skimming and scanning.

They need to learn how to collect data by observation and how to record what they observe, e.g. by drawing up questionnaires and surveys.

Once they have collected data, children need to learn how to analyse it, to judge its value and to draw conclusions, based on the evidence.

They need to learn how to present their conclusions, both orally and in writing, by being provided with opportunities to develop the skills of giving presentational talks and being shown how to structure a written report and taught the appropriate language to use.

Finally, children need to learn how to evaluate their work and to develop the skill of being able to judge critically what they have done well and what areas they need to improve. They can then draw up targets for future work.

Children need to acquire the necessary skills not only to work independently on a topic, but also to work collaboratively as a member of a team. By the end of Key Stage 2 the children should be able to adopt different roles in a group and take responsibility for a variety of tasks, to assist others with their learning, to listen to and respond to what others have to say and to negotiate with others either to reach agreements or to agree to disagree. Therefore, many of the activities in this book involve pair work or group work and there are activities focussing on why group discussion is important and how to have successful discussions.

The aim of this book is to help teachers to make learning active by teaching the necessary study skills that are required for it to take place. The sooner children develop these skills, the sooner their learning becomes more effective.

Activities that develop active learning present children with problems to solve in a way that suggests means of working out the answers rather than just providing them with information and instructing them as to how to find the answers. The role of the teacher is to be a facilitator rather than an instructor, intervening where necessary to point children in the right direction. Children are encouraged to work collaboratively and engage in exploratory talk in group and pair discussions. Thus children are able to develop the skills of how to study through experimenting, exploring and discovering things for themselves.

CHAPTER 1
Planning investigations and projects

This chapter focuses on the skills children need to develop when planning to do a piece of schoolwork. It contains activities and advice on how to plan the investigation of topics across the curriculum, including, e.g. a historical investigation, a design and technology project, a maths problem and a science experiment. There are also suggestions of how to plan the study of books, plays and poems. Children are encouraged to brainstorm ideas, to draw mind maps and flow-charts and to write detailed plans.

PLANNING A PROJECT

Explain that when you are planning an investigation, you will have to decide what information you want to find out, what questions you want answered and how you are going to collect the data you need.

The activities in this chapter are designed to develop children's enquiry skills and their ability to work collaboratively, thinking of relevant questions and carrying out research to collect information.

PLANNING A HISTORICAL INVESTIGATION

Iron Age and Roman Britain
Ask the children to suggest how they would plan to find out about how people lived in a) the Iron Age, b) Roman Britain. What questions would they seek to answer? What evidence could they use to find the answers to their questions? Draw two columns on the board and list the types of evidence they suggest they could find.

Encourage them to think about objects that have been found at archaeological sites and sites that have been excavated that they could visit to find evidence of how the Celts and Romans lived. Mention sites like the forts the Romans built on Hadrian's Wall and Roman villas. Talk about written evidence and make the distinction between primary sources – the books Romans wrote themselves – and secondary sources – books that people have written about how the Romans lived.

Encourage them to think how they could use the internet to find out about both the Celts and the Romans from a website such as www.resourcesforhistory.com and explain that there are apps available to take them on interactive tours of some sites such as the Romans app produced by the Corinium museum at Cirencester https://coriniummuseum.org/romans-app

Britain in the 1950s
Then ask the children in groups to list the evidence they could use to find out about how people lived in Britain in the 1950s.

Encourage them to think of all the things that they could use as evidence about life in Britain in the 1950s that are not available when investigating earlier times, e.g. photographs, films, TV and radio programmes, tape recordings, first-hand accounts from people still living as well as written sources, such as newspapers, diaries, telegrams and postcards, as well as numerous artefacts and buildings. End with a discussion of which sources of evidence would be available if they were investigating Victorian Britain.

PLANNING A DESIGN AND TECHNOLOGY PROJECT

Invite the children in pairs or groups to plan a design and technology project. You can choose a project linked to the topic you are studying. For example, if the topic is the Romans, the children could design and make a model of a Roman amphitheatre or a Roman villa or a model of a Roman soldier. Alternatively, if the topic is Vikings they could design and make a Viking helmet or a Viking hut.

A MODEL OF A ROMAN SOLDIER

Explain the aim of the activity – to plan how to design and make a model of a Roman soldier. Encourage them in groups to make a plan by drawing up a list of questions they will need to answer, such as what did a Roman soldier look like? How can we find out? What materials and tools will we need? How will our model stand up? Invite groups to prepare a detailed plan and to present it to the rest of the class. Discuss how a good plan will include:

- details/pictures of what a Roman soldier should look like;
- design of how your Roman soldier will look and how he will stand up;
- materials you will need;
- tools you will need;
- the method you will use to make him step-by-step;
- any problems you think you may encounter and how you might resolve them.

MAKING A BADGE

Unlike the activity designing a Roman soldier, in which the children have to be historically accurate, this activity is more open-ended. Explain that the activity is to plan how to design and make a badge. The group will have to decide what sort of badge they are going to make by researching different badges. They will have to discuss the purpose of different badges and why people wear them, e.g. to prove identification and for security purposes, to advertise a product, to show that you belong to a team or group, to show your religion or to support a cause or a charity. Having determined the purpose of the badge, encourage the children to look at different designs of badges, their shape and size, how they are fixed on and the different materials they are made of. Encourage them to make two or three alternative designs, then to choose between them, taking into account their suitability for their purpose and whether one of them would prove easier to make. They should then make a detailed plan and present it to the rest of the class.

A GEOGRAPHY PROJECT – OUR ENVIRONMENT

Explain that the purpose of the project is to study the main features of the environment of the village or town where the school is situated and that you are going to go on a walk around the neighbourhood to observe and record the main features.

Put the children in groups and ask them to discuss what they think are the main features they should be observing and how they should record them.

Encourage the groups to think of features such as the landscape, the buildings, the transport facilities, local amenities, workplaces and houses and to draw up a list of things they would look out for.

Ask groups to focus on one particular feature, such as buildings, and to think of things they can look out for during the walk, such as: What different buildings are there? What are they used for? When were they built? What materials are they constructed of?

Encourage the children to think about how they will record their observations and what equipment they will require, such as a camera, clipboards and pencils. Encourage them to design survey sheets.

After the walk, invite them to share their observations and what they found out about the feature they were observing.

A HISTORY PROJECT – THE HISTORY OF THE ENVIRONMENT

As a follow-up project you can invite the children to research local history, e.g. by finding out what the environment was like either 50 or 100 years ago. Ask the children in groups to suggest what evidence they could collect to answer questions, such as: What was the school like? What were the houses like?

In addition to using the internet, encourage them to suggest how they might interview older relatives, friends and neighbours, visit a local museum, contact the local historical society, invite older people to bring in artefacts and photographs and study old maps of the area. List what they suggest on the board. Then ask the groups to plan their research, carry it out and report their findings.

A SCIENCE INVESTIGATION

Explain to the children that if they are going to carry out a scientific investigation, they need to draw up a plan.

First, they will need to decide what questions they want answered. For example, do they want to find out what happens if they make a change to an object or a substance? Do they want to find out how it occurs and when it occurs? Do they want to understand why it occurs?

The flow chart (Figure 1.1) shows a step-by-step plan.

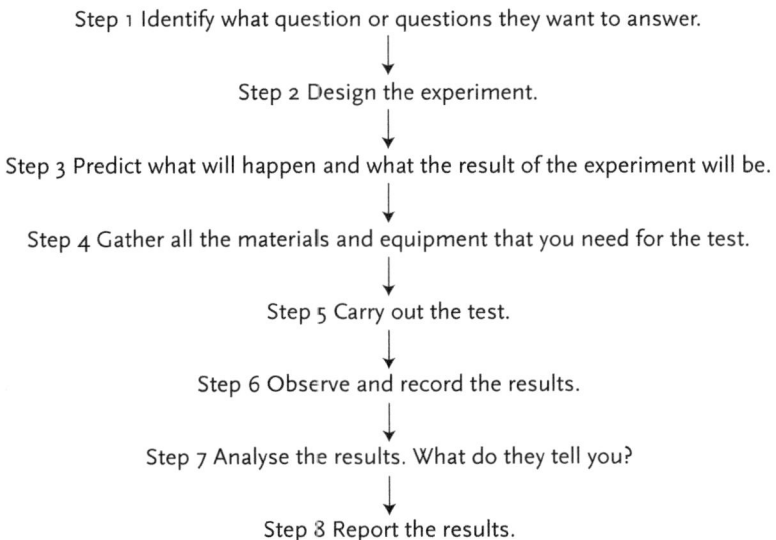

Step 1 Identify what question or questions they want to answer.
↓
Step 2 Design the experiment.
↓
Step 3 Predict what will happen and what the result of the experiment will be.
↓
Step 4 Gather all the materials and equipment that you need for the test.
↓
Step 5 Carry out the test.
↓
Step 6 Observe and record the results.
↓
Step 7 Analyse the results. What do they tell you?
↓
Step 8 Report the results.

Figure 1.1 Flowchart

Plant life and growth

Encourage the children to focus on plants and in groups to think of what questions they would need to find answers to in order to work out what plants need for life and growth. Ask them to suggest what plants need and to think of tests they could carry out to find out if their ideas are correct. They can then share their ideas in a class discussion.

MAKING A MIND MAP

Explain that a mind map or spidergram is a type of diagram which can be used to show different facts and ideas connected to a particular subject or topic. Draw a circle in the centre of the board and write 'How can I find the information I need?'. Ask the children to suggest the various sources of data that can be used to find information on a topic and build up a spidergram of their suggestions. Figure 1.2 is an example of such a spidergram.

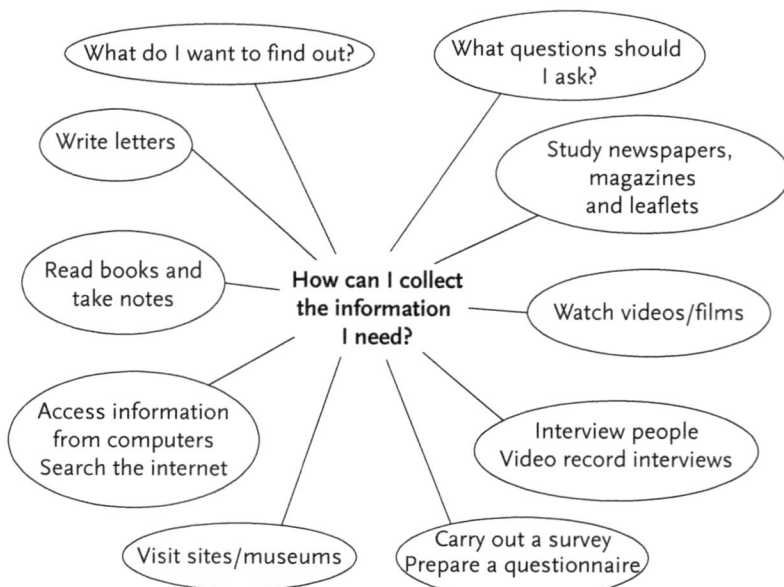

Figure 1.2 A spidergram

Encourage groups to choose one of the investigations listed below, to discuss sources of information that they could use to collect the information they need and to draw a spidergram.

Topics for investigation:

- What life was like for rich children and poor children in Victorian times?
- How the local area has changed during the last 100 years?
- Where deserts are found and what animals and plants are found in them?
- The first motor cars.

You can encourage the children to share their ideas by using apps such as Simple Minds and Popplet which are designed to enable them to develop their mind maps on iPads.

TACKLING A MATHS PROBLEM

Discuss with the children how they plan to tackle the problem.

Ask them what information do they have and what they need to find out or do. What questions do they need to ask?

What operation(s) they are going to use. Will they try to do it mentally, on paper or with a calculator?

What method do they plan to use? Why?

Encourage them to think about any equipment they will need.

How do they plan to keep a record of what they do?

Can they predict or estimate what the answer or result might be? Why do they think that might be the solution?

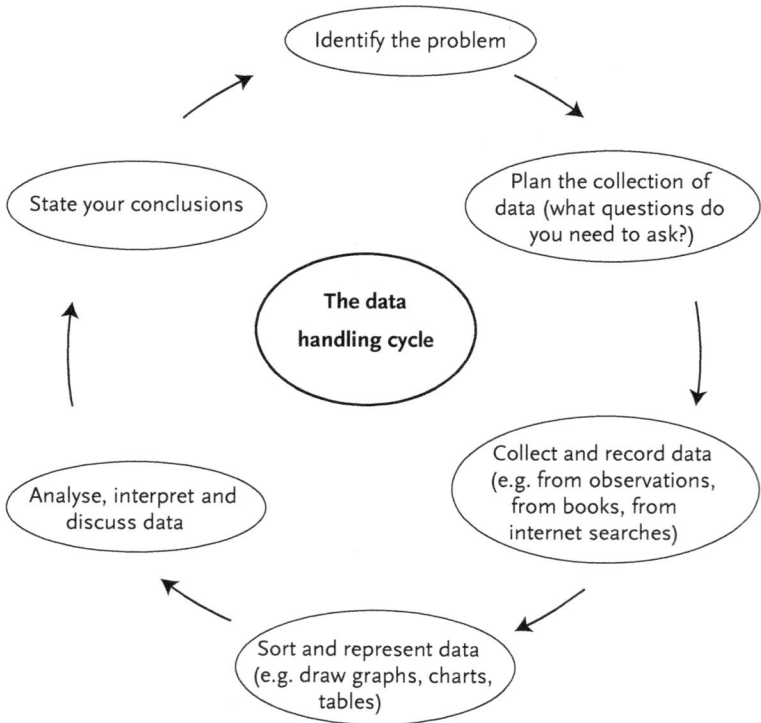

Figure 1.3 The data handling cycle

PROBLEM SOLVING

Explain how they can develop your problem-solving skills by following the data handling cycle (Figure 1.3).

THINK – PAIR – SHARE

This activity involves presenting the children with a problem, e.g. what happens to our bodies when we take exercise and asking individuals to write down what they think happens, why they think it happens and how they could find out whether their ideas are correct. In pairs, ask them to compare their ideas, then to produce a joint statement to share with the rest of the class.

WHAT SHOULD THEY DO?

Present the children with a real-life dilemma in the form of a case study that describes a situation in which a community, family or individual is faced with a problem. For example, there are plans for a new development in the area, such as the building of a wind farm or fracking in the neighbourhood, the construction of a new rail link, a motorway or by-pass or plans for a new runway, or the building of an incinerator. Groups can discuss which people are likely to be for or against the proposal and why. How would they try to persuade others to share their views?

After an initial discussion of the issue, the class can be divided into two groups – those who are for the proposal and those who are against. They can work together to formulate their arguments. The different views can then be presented either in a formal debate or a role play of a public meeting.

As a follow-up activity, groups can be asked to discuss how they might organise an action group to protest against the proposal and what actions they might take. Ask groups to share their ideas in a class discussion and rank the actions in order of which you think would be the most effective.

INVESTIGATING A RIVER

Encourage the children to plan the study of a local river by researching it and then visiting it to find out about it by observation. Brainstorm how they might research it prior to the visit by studying maps and photographs to find out where its source is and whether it is a tributary or has a mouth where it enters the sea. Encourage them to ask such questions as:

- Are there any local issues connected to the river, such as flooding or pollution?
- What use is made of the river? Are there any boats/canoes/rowing boats or pleasure cruisers on it?
- Are stretches of it used for fishing?

Ask them to suggest how they could find the answers to their questions.

PLANNING A BOOK REPORT

Encourage the children to plan a report on a non-fiction book. Talk about what the report should include:

- The title, author and date of publication.
- The name of the series (if it is in a series).
- A summary of the contents – encourage them to look at the introduction and blurb, which will help to give them an overview of the book, in addition to the contents list and chapter headings.
- Details of particular features that encapsulate what the book is about, including parts of the book that they found fascinating.
- A comment about any illustrations, photographs or diagrams.
- A concluding comment on who you think the book is for and to whom you would recommend it.

PLANNING A PLAY AREA

Present the children with a creative matrix and invite groups to come up with a plan for a new play area that is suitable for a particular age group.

Type	A soft play area	A playground with slides	An adventure playground	A woodland trail	A skateboard park
Who for?	Teenagers	5–7 year olds	Babies and toddlers	3–5 year olds	8–12 year olds
Where?	A disused factory	Derelict land	A municipal park	A shopping centre	A new estate

Figure 1.4 Creative matrix: planning a play area

PLANNING A SCHEDULE

Explain that when you are planning a project it is necessary to know how long you have to complete the project and to work out a schedule. Many projects are not as successful as they could have been if the group had made better use of its time.

Put a schedule for the badge-making activity on the board and discuss it. Encourage them to make a similar schedule for a design project.

Week 1 Research different purposes. Decide purpose.

Week 2 Look at different designs. Produce alternatives.

Week 3 Choose design. Collect materials and equipment.

Weeks 4 and 5 Manufacture badge.

Week 6 Presentation and review.

CHAPTER 2

Collecting data – observations, surveys and interviews

Talk with the class about the various ways that you can collect information, which you use will depend upon the nature of the project. For example, if you are studying a habitat you can make observations, recording what you see; if you are studying Roman Britain you may be able to visit the site of a Roman fort or the ruins of a Roman villa or use an app to do a virtual tour, e.g. there is an app which allows you to accompany a Roman soldier visiting the Roman sites in the Brecon National Park. If you wish to find out people's attitudes on an issue you may decide to use a questionnaire. This chapter explores ways of collecting data and analysing it.

OBSERVATION

Data can also be collected by observation. For example, the children can carry out a pond-dipping activity to discover what creatures they can find in a pond. They will need to record their findings. Encourage them to prepare a sheet on which to record their findings. Give them copies of the sheet in Figure 2.1 and discuss how it includes details of the pond, the date and the person doing the observation as well as the creatures which were observed.

```
┌─────────────────────────────────────────────────────────────┐
│                                                             │
│   Pond  dipping        Observer  ...........................  │
│                                                             │
│                                                             │
│   Situation of pond.......................................... . │
│                                                             │
│  ┌──────────────┬──────────────────────────────────────────┐│
│  │ Date         │ Creatures observed                       ││
│  ├──────────────┼──────────────────────────────────────────┤│
│  │              │                                          ││
│  │              │                                          ││
│  │              │                                          ││
│  │              │                                          ││
│  │              │                                          ││
│  │              │                                          ││
│  │              │                                          ││
│  │              │                                          ││
│  │              │                                          ││
│  └──────────────┴──────────────────────────────────────────┘│
│                                                             │
└─────────────────────────────────────────────────────────────┘
```

Figure 2.1 A pond-dipping survey

A WEEKLY WEATHER CHART

Invite the children in pairs to design an observation sheet to record the weather twice a day for a week. Discuss what facts about the weather you want to record, e.g. the temperature, and decide what your weather chart will look like. Then encourage them to compare their charts and to modify their own charts, if they think they could improve them.

As necessary, prompt them to discuss the instruments they will need and what else they should record as well as the temperature – the wind direction and speed, the cloud cover and the rainfall.

As a further activity, they could record the weather forecasts for the area during the week and compare them with their findings.

A LOCAL RIVER

Either look at the plans the children made for the study of a local river (see page 11) or ask them to draw up such a plan.

Then, organise a field trip to observe the river. Invite groups to make a list of what they want to find out from their observations. Here is a list which one group made:

- Are there any locks or weirs?
- How fast is the current?
- Is the river dangerous?
- How steep is the river bank?
- Are there any signs of pollution?
- How is the river used?
- Are there any boats? Are they homes? Leisure cruisers? Barges?
- Is there any sign that the river floods?
- What is the ground like? Is it sandy, rocky or chalky?
- What trees and plants grow on the riverbank?
- Is there any wildlife on the riverbank or on/in the water?
- What fish are found in the water? Are there any fishermen or notices about fishing?
- Are there any special features such as meanders?

Figure 2.2 Investigating a river

Ask the children to plan for the trip by drawing up an observation sheet on which to record their observations.

Figure 2.3 is an example of such a sheet.

Observer's name:	Date of observation:
Name of river	
Location where observation took place	
Are there any river features such as meanders, erosion, signs of flooding? Record details.	
How deep is the river?	
How wide is the river?	
How strong is the current?	
Is there any evidence of human activity to control the river, such as flood defences, weirs or locks?	
What use is made of the river? Is there any evidence of its use by local industry or for pleasure, e.g. for boating holidays or for fishing or rowing?	
Are there any signs of pollution?	
What trees/plants are growing on the river bank? What wildlife mini-beasts, animals, birds - or signs of wildlife are there?	
Are there any further interesting facts that you observed about the river?	

Figure 2.3 A river observation sheet

RECORDING YOUR OBSERVATIONS – HELPFUL HINTS

Explain that whenever you collect data by observing what happens it is important to keep a record of your observations.

Put this list of tips on the board. Discuss the advice that is given and ask the children in pairs to select what they consider to be the two most helpful hints.

- Use a clipboard so that you can press on something when you are writing.
- Use a bound notebook whenever possible to make sure that pages are not lost.
- If you are using charts on single sheets of paper, put completed sheets inside a folder.
- Write neatly, so that you can read what you have written.
- Don't rely on remembering what you observed. Always keep a written record.
- Make sure you put the date on each entry you make.
- Put a title. For example, you can use the title of the experiment you are doing.
- Put your observations in a logical order.
- Check that you have recorded everything you planned to record.
- Get a partner to glance at your notes to see if they are as detailed as necessary.
- Alternatively, record your observations on a tablet or iPad.

CARRYING OUT A SURVEY

Explain how information can be collected by carrying out a survey.

HOW DO YOU COME TO SCHOOL?

Ask the children to design a survey sheet to find out how children travel to school. Explain that first they will need to list all the different ways that they may travel to school. Then they will need to decide how they would take into account the fact that sometimes they may use a different method. How could they show this?

Encourage them to share their ideas with other groups and to agree on a sheet they will use.

Ways of coming to school	Usually	Sometimes	Very occasionally
School bus			
Walk			
Scooter			
Car			
Taxi			
Minibus			
Bicycle			
Public transport (train)			
Public transport (bus)			

Figure 2.4 Survey sheet: ways of coming to school

Note: If necessary, provide them with the model in Figure 2.4 to use:

Ask the class to fill in the sheet by passing it round the class and getting each person to put a tick in the appropriate column. What does the survey tell them? Encourage them to give the survey to another class to complete and to compare the results. Are the results the same or different?

FACTS OR OPINIONS

Stress that it is important to have a clear idea of the information that the survey is intended to find out. For example, do they want to establish facts or to explore what people's opinions are?

Show them the two examples in Figures 2.5 and 2.6 and discuss how the first example is designed to collect facts about a hedgerow, while the second is designed to find out what people's attitudes to litter are. (Note: the children could then use the sheets to carry out the surveys.)

Invite the children to use one of these two surveys as a model.

For example, they could carry out a survey of the local park to establish the facts about what facilities it provides or a survey of the local market.

Alternatively, they could carry out a survey of people's opinions of how safe they think the local area is or of public transport in the area where they live.

SEMI-STRUCTURED INTERVIEWS

Explain that interviews are a useful source of information, both for establishing facts and for finding out opinions. An interview can be useful, e.g. if you want to discover how something has changed over a period of time and how that change has affected people's lives.

HOW SCHOOLS HAVE CHANGED

Encourage the children to investigate how schools have changed by interviewing an older person, such as a relative or friend of the family about what their primary school was like, what lessons they had, what uniform (if any) they wore and what the classroom was like.

Ask the children in pairs to discuss what information they want to find out and the questions that they want to ask. Point out the difference between a closed question, such as 'Did you get lots of homework?' and an open question such as 'What homework did you get?'

A hedgerow survey

1. Is the hedgerow straight? Yes No

2. Is the hedgerow a) very tidy and clipped?
                         b) ragged and straggly? Put a circle round your answer.

3. What trees and shrubs are there? Circle those you saw and add the name(s) of any others.
   hawthorn   blackthorn   holly   hazel
   ash   chestnut   oak   elm   maple   elder

Others_____

4. What plants and flowers are there? Circle those you saw and add the names of any others
   ivy bramble old man's beard honeysuckle hedge bindweed hogweed
   nettle moss dandelion cow parsley thistle

Others_____

5. List any birds you saw _____

   _____

   _____

6. List any creatures you saw or signs of creatures (e.g. molehills, burrows)

7. How old do you think the hedgerow is?
   Why?_____
   _____
   _____

8. Any other useful information you would like to add:

Figure 2.5 A hedgerow survey

A Litter Survey

Age: Under 18   18–65   over 65

Read each of these statements and put a tick in one of the boxes to say whether you agree with it, disagree or are unsure.

People make too much fuss about litter.	Agree
	Disagree
	Not sure
People who drop litter should be given on-the-spot fines.	Agree
	Disagree
	Not sure
You should always dispose of litter in litter bins.	Agree
	Disagree
	Not sure
There would be less litter, if there were more litter bins.	Agree
	Disagree
	Not sure
It's mainly children and young people who drop litter.	Agree
	Disagree
	Not sure
If there was less packaging, there would be less litter.	Agree
	Disagree
	Not sure
We need to have litter wardens, with powers to issue people with tickets like traffic wardens.	Agree
	Disagree
	Not sure
Fast food outlets should be responsible for keeping pavements near them free of litter.	Agree
	Disagree
	Not sure
The council should employ more street cleaners to pick up litter.	Agree
	Disagree
	Not sure

Figure 2.6 A litter survey

CLOSED QUESTIONS AND OPEN-ENDED QUESTIONS

Explain that closed questions are questions that restrict the answer to one of a list of choices. For example:

Which of these creatures did you observe in the farmer's field? Horse   cow   sheep.

Note: The question restricts the answer to the three animals it lists. It doesn't, include e.g. pigs, turkeys or geese.

Another type of closed question is a question that only requires a yes or no answer e.g. 'Do you eat five portions of fruit and/or vegetables each day?'

The advantage of closed questions is that they provide data that is easy to analyse.

Explain that an open-ended question does not give any choices, so there is no restriction on the answer.

TYPES OF QUESTION

Children in pairs have to decide which of these questions about schools are closed questions and which are open questions. Then to identify which of the questions are likely to produce the most information.

1   How do you travel to school?
2   What time does school start?
3   Do you have an assembly every day?
4   Which lessons do you like best?
5   Are you learning to play an instrument?
6   What do you do at break-time?
7   Do you have a packed lunch?
8   What uniform do you have to wear?
9   What things would you like to change about school?

INVESTIGATING TRAFFIC IN YOUR LOCAL HIGH STREET

Encourage the children to observe the traffic through the local high street to ascertain how safe it is and to discover what local residents think about traffic through the town and whether it ought to be restricted or banned in any way.

Invite them to prepare an observation sheet and to count the number of different vehicles they see passing through it. Get them to count the number of cars, vans, lorries, taxis, buses, emergency vehicles, motorcycles, bicycles during a five minute period passing a particular spot at a particular time. They could include space on the sheet for them to say whether the traffic moved slowly, flowed steadily or drove fast. Did they observe any pedestrians trying to cross the high street? Was it easy and safe to cross, or hard and dangerous?

Encourage them to draw up a survey for local residents to complete giving their views on traffic through the high street. Point out that they should not just ask closed questions such as:

- Do you think the high street is safe for pedestrians?
- Should all traffic be banned from the high street from 6 am to 6 pm?

They should also include questions such as:

- What do you think should be done about traffic through the high street?
- Who do you think would be in favour of a scheme to close the high street to traffic?
- Who would be likely to oppose such a scheme?

Note: They could present their observations of the traffic in the form of a bar chart and hold a class discussion of what they found out from their surveys.

RANKED RESPONSES

It can be useful when you are seeking to find out people's opinions on a topic to ask them a question in which they rank a number of statements.

WHAT SHOULD A SCHOOL AIM TO DO?

Talk about how you can find out what people think about what the aims of a school should be by asking them to rank the a number of views in order of importance:

To prepare children for a job.

To teach literacy and numeracy.

To teach people to be tolerant of different faiths.

To teach moral values.

To teach children about the past.

To give children the opportunity to make friends.

To teach children about the world they live in.

To teach children to become good citizens.

To help children to discover their talents.

To give everyone an equal chance to succeed.

To teach you how to use technology.

Put the list of aims on the board and encourage the children to rank the aims in order of importance 1 being the most important and 10 the least important. Talk about how this activity provides data that can be collated and analysed easily.

DO YOU EAT A HEALTHY DIET?

This activity is for two classes. First, ask the children in groups to discuss what constitutes a healthy diet, then in pairs to draw up a series of questions to find out whether a person eats a healthy diet. Invite the children to join up with a partner from another class and to interview them. After the interview ask them to decide whether or not the person they interviewed eats a healthy diet.

Put the flow-chart in Figure 2.7 on the board and explain that the data you collect will depend on the questions you ask.

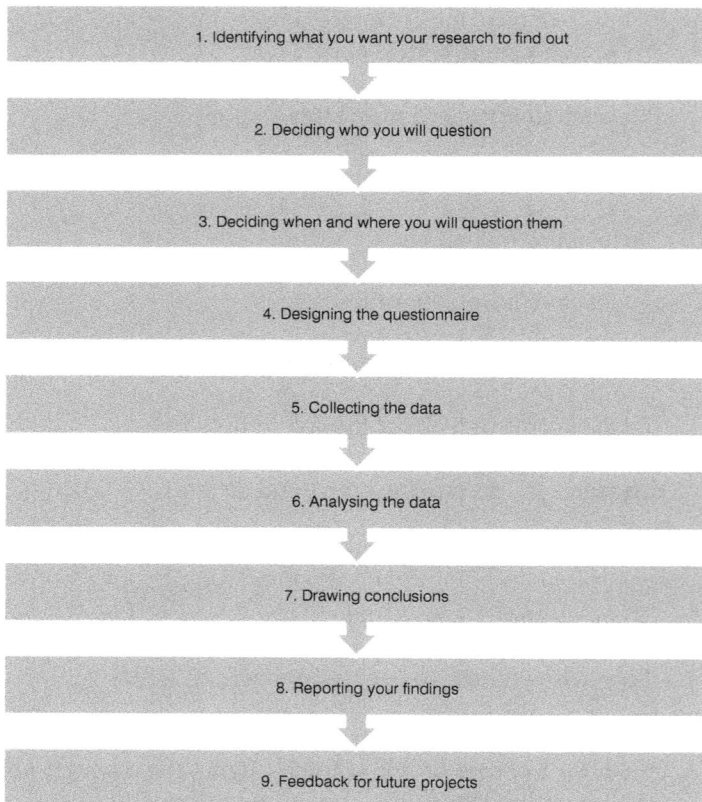

1. Identifying what you want your research to find out

2. Deciding who you will question

3. Deciding when and where you will question them

4. Designing the questionnaire

5. Collecting the data

6. Analysing the data

7. Drawing conclusions

8. Reporting your findings

9. Feedback for future projects

Figure 2.7 The process of collecting, recording and reporting data

CHAPTER 3

Developing reading skills

This chapter focuses on developing children's ability to find information from books and on developing their reading skills.

FINDING INFORMATION – USING AN INDEX

Explain that an index lists the subjects that are included in the book and the pages on which the subjects can be found.

Put this index from a book called 'Discovering Dinosaurs' on the board and invite the children in pairs to answer the following quiz:

On which page or pages can you find out about:

1   How the dinosaurs died out.
2   How the dinosaurs reproduced.
3   Digging out dinosaur bones.
4   Different kinds of dinosaurs.
5   Where dinosaurs have been found.
6   Plant-eating dinosaurs.
7   Meat-eating dinosaurs.
8   How clever dinosaurs were.
9   What colour dinosaurs were.
10 How big the dinosaurs were.

Answers: 1. p.26; 2. p.21; 3. pp.6–9; 4. pp.10–19; 5. pp. 2, 4–7; 6. p.10; 7. p.10; 8. pp.24, 18–19; 9. p.20; 10. p.3.

INDEX

INDEX

Allosaurus 13

Apatosaurus 10

bird-hipped 10, 16–19

bone-headed 17

brains 18,19

breeding 21

Brontosaurus 12

carnivorous 10

Diplodocus 11

discoveries 2, 4–7

duck-billed 16

excavation 6–9

extinction 22

fossils 4–7

herbivorous 10

Iguanadon 6,9,16

lizard-hipped 10–15

sauropods 10–12

size 3

skin 20

Stegosaurus 18

Theropods 10,13

Triceratops 19

Tyrannosaurus rex 14–15

USING THE CONTENTS LIST

Put the contents list from a book 'All you need to know about Pirates' on the board and invite the children to answer the following quiz: On which pages would you find information about: 1. Blackbeard; 2. cutlass; 3. scurvy; 4. crow's nest; 5. Roman pirates; 6. Robert Louis Stevenson's 'Treasure Island'; 7. the meaning of the word 'maroon'; 8. gold doubloons; 9. Anne Bonny; 10. pirates in the West Indies?

Answers: 1. pp. 16, 26; 2. p. 22; 3. p. 20; 4. p.18; 5. p.8; 6. p.32; 7. p.38; 8. p.38; 9. p.28; 10. p.16.

CONTENTS

Who were the pirates?	6
Pirates in the Ancient World	8
Viking raiders	10
The corsairs of Malta	12
Privateers of the Spanish Main	14
Pirates of the Caribbean	16
Pirate ships	18
Life on board a pirate ship	20
Pirate weapons	22
Pirate treasure	24
Famous pirates	26
Women pirates	28
Desert islands	30
Pirates in literature	32
Pirates in films	34
Pirates in the modern world	36
Glossary	38

DEVELOPING YOUR READING SKILLS

Explain that there are many ways of reading a passage to gather information. The method you use will depend on the purpose of your reading and the type of material you are reading.

There are three main methods you can use:

1 **Skimming.** You read quickly, not trying to take in all the details, in order to get a general idea of what a passage or article is about or to find out what information a book contains. If you are skimming through a book you will look at such things as the contents, introduction and chapter headings, glossary and index.
   Your aim is to find out whether the book contains the information you need. Similarly if you are skimming through an article or a passage, you won't try to take in everything it says at once. Your aim is to find out whether it interests you and contains useful information that you are searching for.

2 **Scanning.** You run your eye quickly down the page, not trying to read every sentence, searching for the sentence or paragraph that will give you a particular piece of information. Scanning is a useful method when you want to check on an important point, such as the meaning of a particular word or the date of an event or to extract some particular details from a passage.

3 **Close reading.** You read a chapter, passage or article carefully, studying it in detail in order to obtain as much information as possible from it. Your aim is to try to understand all the facts and ideas it contains, so you may read it several times and take notes in order to help you to remember what you read.

IN THE NEWS

This is an activity for pairs. Give each pair a copy of an old newspaper. Ask the pairs to skim through the paper, looking at the headlines to find out what are the three main news stories in that paper.

Encourage them to choose one of the main stories to read closely. Then, ask them to take it in turns to retell the story without looking at the paper. Get them to check whether they left out any important points by scanning through the article.

SPOT THE KEYWORDS

This is a game for two teams. Divide the class into two teams. Give out copies of a piece of writing consisting of several paragraphs about the topic which you are studying. Read the passage to the children and ask the children individually to use a highlighter to identify the keywords in the passage.

Each person then exchanges their highlighted copy with a member of the other team. The teacher goes through the passage and discusses which words are the keywords. A point is awarded for every keyword that has been highlighted and a point is deducted if a word has been highlighted that is not a keyword. The team whose members score the most points is the winner.

THE FIVE Ws

Explain how journalists are taught to include answers in their reports to the five W questions – Who? What? When? Where? Why? Present the children with a text about a historical event, such as the Great Fire of London, and ask them to read it in order to find the answers to the five Ws. Then ask the children to draw five columns labelled Who? What? When? Where? Why? and to make notes of what they have learned.

UNDERSTANDING CHARTS, TABLES AND GRAPHS

Explain that information can be presented in the form of charts, tables and graphs and that it is important to be able to interpret data that is presented in this way.

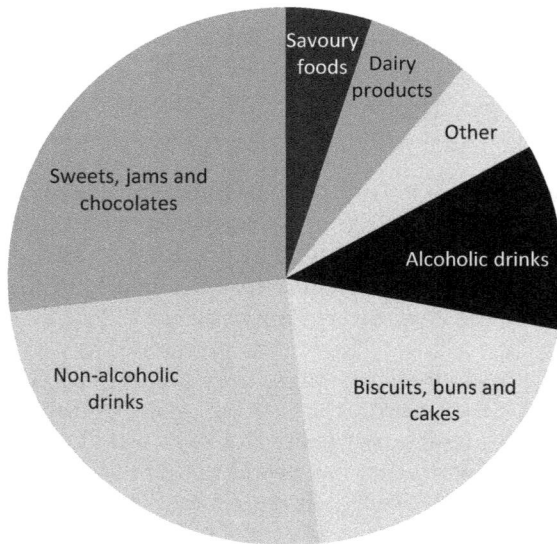

Figure 3.1 Sources of added sugar in our diets

PIE CHARTS

A pie chart looks like a pie that has been cut into slices. Pie charts are used to show the proportions of a set of figures.

Put Figure 3.1 on the board. Ask the children what it tells them about the sources of added sugar in our diets and to suggest two things you could do to cut down the amount of sugar in your diet.

BAR CHARTS

Explain that a bar chart is a chart that shows information in the form of rectangular bars of different heights. Figure 3.2 shows how many children in a class have pets and the types of pets they have. The horizontal axis shows the different types of pets and the vertical axis shows the number of children who have that type of pet. What do you learn about children and their pets from the chart?

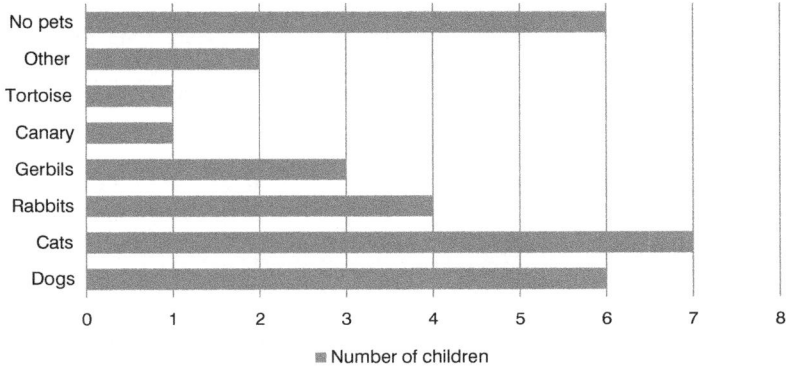

Figure 3.2 A bar chart

PICTOGRAMS

Explain that a pictogram is like a bar chart which has pictures in it. Demonstrate what a pictogram looks like by drawing animals in the bar charts the children drew, so that column one contains six dogs, column two contains seven cats and so on.

HANDLING DATA FROM TABLES

Put the table in Figure 3.3 'What children read' and Figure 3.4 'MPs elected at 2010 and 2015 general elections' on the board. Explain that it represents the findings of a survey in which 24 children of each age were asked what they preferred to read. Then ask the children in groups to say what they can learn from it about children's reading habits at different ages.

Ask the children to study the table in Figure 3.4 which gives the numbers of people from different parties who were elected to the House of Commons in the general elections of 2010 and 2015. What do they learn from it about how the composition of Parliament was different in 2015 from in 2010? Ask the children to share what they learned and point out that while the table shows the results of the election, it does not show why the results were different.

What children read	Picture books	Chapter books	Newspapers	Magazines	Comics
Age 11	0	18	2	3	1
Age 10	0	16	1	3	4
Age 9	1	17	0	2	4
Age 8	1	16	1	2	4
Age 7	2	16	0	2	4
Age 6	6	10	0	2	6
Age 5	8	8	0	1	7

Figure 3.3 What children read

Number of MPs elected from each party	2010 election	2015 election
Conservatives	306	330
Labour	258	232
Liberal Democrat	57	8
Scottish Nationalist	6	56
Democratic Unionist Party	8	8
Independent	1	1
Sinn Fein	5	4
Plaid Cymru	3	3
Social Democratic and Labour Party	3	3
Ulster Unionist Party	0	2
Green Party	1	1
Alliance	1	0
UK Independence Party	0	1
Speaker	1	1

Figure 3.4 MPs elected at 2010 and 2015 general elections

LINE GRAPHS

Explain that a line graph, like a bar chart has an x-axis, which is the horizontal line and a y-axis, which is the vertical line. The graph (Figure 3.5) shows how the world population has increased decade by decade since 1950 and how it is predicted to increase up to 2050 with the x-axis showing the years in decades and the y-axis showing the size of the population in billions.

Put the graph on the board and ask the children what they can learn from the graph about how the population has grown since 1950.

Point out that these figures do not answer the question why the population has grown.

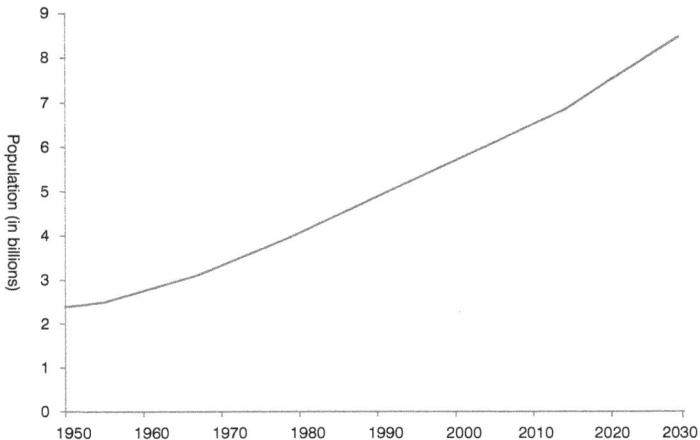

Figure 3.5 World population growth 1950–2030

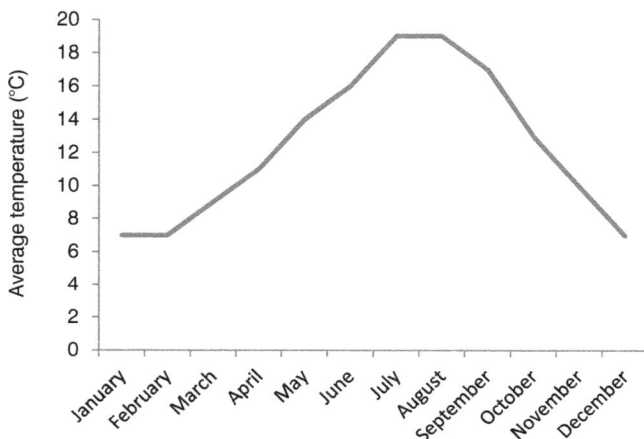

Figure 3.6 Average monthly temperatures, London, UK

Put the line graph (Figure 3.6) on the board. It shows the average annual temperature for each month of the year in London. Discuss how the x-axis shows the months and how the y-axis shows the average temperature in Celsius.

Encourage the children to use the figures (below) which state the average annual temperature in Celsius month by month for the Australian city of Sydney to draw a graph similar to Figure 3.6. Then ask them to compare the two graphs and to discuss what they learn from comparing them.

January 23°C; February 23°C; March 22°C; April 19°C; May 16°C; June 14°C; July 13°C; August 14°C; September 16°C; October 18°C; November 20°C; December 22°C.

CHAPTER 4

Developing note-making skills

MAKING NOTES

Explain that good note-making is an important skill because it not only keeps a record of the information you have found out, but also because it helps you to think about what you have learned.

SNAKES

Contrary to what many people think, a snake's skin is dry rather than slimy. Although the skin of a snake is covered in scales, it is smooth.

Although some snakes, such as cobras, kill their prey by biting them and injecting them with poison, that isn't what most snakes do. Most snakes capture their prey in their jaws and then eat them by swallowing them whole. Large snakes, such as pythons, will strangle their prey then swallow them.

Snakes are cold-blooded reptiles which cannot regulate the temperature of their bodies as warm-bloodied creatures can. So, snakes do not like the cold. When it gets cold in winter, some snakes will go underground and hibernate.

IDENTIFYING KEYWORDS

Make copies of the three paragraphs about snakes and ask the children to identify the keywords and phrases in each paragraph and to highlight them. Then discuss which words and phrases they have highlighted.

In order to make good notes, children need to learn the skill of identifying the keywords and phrases in the texts they read.

SUMMARISING

Explain that learning how to make a summary is a skill that will help children to develop their note-making skills, as it involves identifying the main points of a paragraph or number of paragraphs.

A step-by-step approach to making a summary:

1   Skim through the paragraph to get an idea of what the text is about.
2   Read it through slowly highlighting keywords and phrases by underlining them, or if it is a digital text marking them by using bold, italics or a different size font.
3   Cross out or delete any phrases, clauses or sentences which you can ignore, such as examples or unnecessary elaborations.
4   Think of a topic sentence which you can use as the start of your summary.
5   As far as possible write the summary in your own words.

Explain that a good summary:

•   is much shorter than the original paragraph
•   includes all the main points
•   is in your own words and is not just a copy of the paragraph.

Give the children a copy of this paragraph about deserts and ask them to highlight the keywords and phrases in it and to cross out any details they can ignore. Then discuss the words and phrases they have highlighted and what they have crossed out. In the discussion make sure they understand how the second sentence can be ignored as can the sentence about frying an egg. Then ask them to write a summary of the paragraph.

Deserts are found in areas of the world where the air is very hot and there is very little rainfall. In the hot, dry deserts the sun blazes down from a clear sky. During the day, the ground gets very hot. The temperature on the ground can reach over 75 degrees C. The rocks can get so hot that you could fry an egg on them! But at night the temperature falls rapidly. The temperature of the air can fall from over 50 degrees C to 20 degrees C.

NOTE-MAKING

Explain that there are several different ways of note-making and that you should choose the one that you find works best for you.

LINEAR NOTES

This is the note-making method most people use. The notes are written down the page, using headings and subheading.

TWO COLUMN LINEAR NOTES

Explain that one common way of making notes is to divide a piece of paper into two columns with the left hand column taking up one third of the page and the right hand column two thirds of the page. You write the title at the top of the left hand column and use it for chapter headings, sub-headings and keywords. In the right hand

Norman Castles	
Sites	On hills. Diff. to attack.
	Nr water e.g. river crossing enemy needs 2 cross.
	Water supply during siege
	Water 4 moat
Motte	Huge mound of earth w. wooden tower on top.
	Fence round motte - palisade
Bailey	Large outer area. Stables etc.
	Built of sticks and clay - wattle and daub
	M & B Cs cd be built quickly e.g. York 1068
Keep	Later Cs had stone towers.
	Ks had stone walls kn as curtain walls
Battlements	Round roof of castle w.spaces 4 archers.

Figure 4.1 An example of two column linear notes

column you write any information that you will need to remember such as definitions, explanations, evidence, facts and opinions.

Put Sandra's notes on Norman castles (Figure 4.1) up on the board. Discuss with the class how she has put her headings in the left hand column and the details she wants to remember in the right hand column.

Ask the children to pick out abbreviations she uses. Which ones are commonly used? For example, nr for near, w. for with. Which has she made up herself? Draw attention to diff for difficult, Cs for castles and cd for could.

SPIDERGRAMS

Explain that another widely used method of making notes is to draw a spidergram, which many people find a good way of

Supply during siege   Diff. to attack

For moat———Near water   On hills

Huge mound

e.g. river crossing   Sites   Motte ——— wooden tower

palisade (wooden fence rd motte)

NORMAN CASTLES

stables etc.

Keep   Bailey

wattle and daub

Stone tower   curtain stone walls

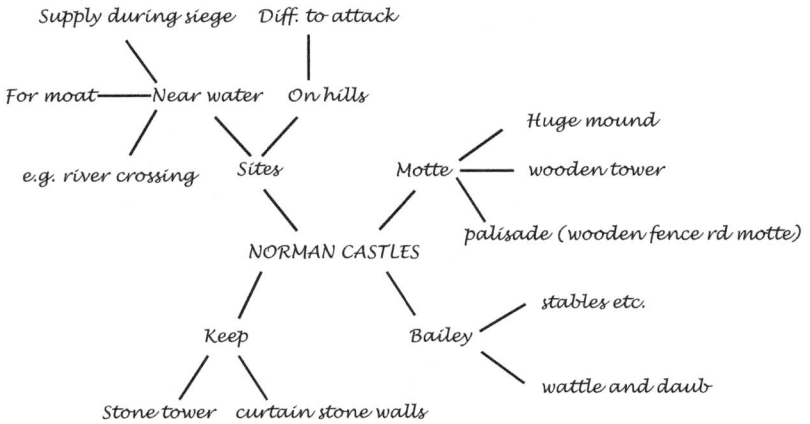

Figure 4.2 A spidergram

organising the facts and ideas about a topic. Spidergrams are often referred to as mind maps.

To make a spidergram you write the name of the topic in the centre of the page. You put each of the main points as a branch radiating out from the middle. Then you add further branches of the details and examples you need to note about each main point.

Draw the spidergram (Figure 4.2) on the board step-by-step showing how Sandra would have developed a spidergram, if she had chosen to use one rather than to make linear notes.

Encourage individuals to make notes on a passage related to the topic they are studying by drawing a spidergram.

Compare their spidergrams in a class discussion, pointing out that the advantages of spidergrams include:

- You are able to pick out the main points at a glance.
- You are less likely to include irrelevant details, because the notes are all on one page, so you have to be precise.

ACTIVE NOTE-MAKING

Talk about how good note-making is active rather than passive. Ask individuals to draw two columns on a sheet – one labelled 'Active note-making' and the other 'Passive note-making' – then put the list of activities (below) on the board and ask the class to decide which are active methods of note-making and which are passive and to list them in the appropriate column.

- Writing notes on everything you read.
- Looking out for relevant material that answers your questions.
- Being clear about what you want to learn from your research before you start.
- Accepting that any source that gives you information is worth taking notes from.
- Ignoring unnecessary details.
- Copying out a passage word for word.
- Noting down facts without thinking about how to organise them.
- Distinguishing between fact and opinion in your notes.
- Evaluating the reliability of the source.
- Looking for evidence to note that backs up an opinion.
- Printing out what you find on the internet without thinking about its relevance.
- Using headings and sub-headings to organise what you have learned.
- Being selective about what you note down.
- Cutting and pasting passages without making notes on them.
- Going over your notes to see that they make sense.

When they have finished, hold a class discussion about what constitutes active note-making. Then invite the children in pairs to role play a scene in which Active Andrew explains to Passive Paula what is the difference between active and passive note-making and why it is important for Paula to change her habits.

USING ABBREVIATIONS

Discuss how using abbreviations makes it possible to keep notes shorter.

Instead of writing 'for example' you can write e.g. similarly, you can write EU instead of European Union.

Everyone uses abbreviations such as rd for road, st for street, v for very, w for with, no for number, Mt for mountain, C20th for twentieth century, K for King and Q for Queen.

Point out that it can be useful to make up abbreviations of your own, particularly for words that occur frequently when you are studying a particular topic. For example, when studying the Vikings, you could use V or Vik for Viking, inv for invasion/invader and l/ship for longship. Or when making notes on the Battle of Hastings, you could use b of Hast for the battle of Hastings, WtC for William the Conqueror, N for Norman and so on.

WAYS TO ABBREVIATE WORDS

Discuss how you can shorten almost any word to make note-making easier.

1   Use only the beginning of the word.

approx for approximately
Eng for England
esp for especially
lang for language
min for minimum
max for maximum
para for paragraph
Parl for Parliament
Scot for Scotland
temp for temperature

2 Leave out letters. (Note often a vowel or vowels are among the letters left out.)

bk for book
dept for department
E for east
lft for left
mkt for market
N for north
nr for near
pt for point
rlwy for railway
rt for right
S for south
St for Saint
W for west
yr for year

3 Use only the initial letters of names of countries and organisations.

BBC for British Broadcasting Corporation
USA for United States of America
EU for European Union
NATO for North Atlantic Treaty Organisation
UK for United Kingdom
UN for United Nations
TUC for Trade Union Congress

THE ABBREVIATIONS CHALLENGE

Inform groups that they have five minutes to list as many abbreviations as they can think of, which are made by using only the beginning of the word. You can repeat the activity by giving the children five minutes to list abbreviations that are made by missing out vowels.

AN ABBREVIATIONS CHART

Give out post-it notes and invite the class to write useful abbreviations on them and stick them on an abbreviations chart. Encourage them to write abbreviations that can be used for the notes they are making on the topic that you are doing with them.

USING NUMBERS AND SYMBOLS

Discuss how you should use numbers, such as 3 instead of writing 'three' when writing notes and how symbols can be useful too. Ask them in groups to make a list of common symbols that they could use. Then to compare them with a list you put on the board. Your list could include:

& % @ $ £ = < > +

MATCHING SYMBOLS

This is a game for pairs. First, they need to make two sets of cards. On one set of cards they put a symbol on each card and on the other set of cards they put its meaning.

Then they shuffle the two sets of cards and lay them face down on a table. The players take it in turns to turn over two cards. If one shows a symbol and the other its meaning then the player picks up the two cards and has another go. If the two cards do not make a match, then they are turned back face down on the table and the other player has a go. The game continues until all the cards have been picked up. The winner is the player with the most cards.

THE FIVE Rs STRATEGY

Explain that making good notes is part of the process of learning what you need to know, but it does not necessarily mean that you have learned what you set out to learn. It is important to adopt a learning strategy that involves thinking about what you have learned and reflecting on the meaning of the facts, ideas and concepts that you have recorded and periodically returning to the notes to review what you have learned.

A useful strategy to adopt is to follow the five Rs of note-making.

Step 1 Recording. This involves making written notes of the main facts, ideas and concepts, identifying keywords, making sure that you have the answers to the questions that you want answered.

Step 2 Reducing. This means that you reduce the notes by picking out the key points and summarising what you need to know should you be tested on the topic. Often it involves transferring your notes onto index cards.

Step 3 Reciting. Check to see that you have used your own words rather than simply copied out sentences from the text. Reciting what you want to remember in this way will help you to understand it better.

Step 4 Reflecting. Think about what you have learned – the concepts, their meaning and the implications. Unless you reflect on your learning, you may fall into the trap of repeating it parrot-fashion without any real understanding.

Step 5 Reviewing. If you review your learning at regular intervals, you will keep it fresh in your mind.

REFLECT AND REVIEW

Encourage the children to choose a set of notes they made recently and to use the five Rs strategy to reinforce what they have learned from taking notes on this subject. Ask them to join up with a partner and to explain what they learned. They can repeat this activity using different sets of notes, until the strategy becomes a habit.

TAKE MY ADVICE

Encourage the children to do a role play in which Trisha Topclass offers advice to Margery Muddlednotes on how to store your notes so that you can find them easily. Ask some of the pairs to perform their role plays to the whole class, then lead a class discussion in which you ask why it is important to be able to find notes easily and what is the best way to store them.

Points to make in the discussion are:

- You can waste a lot of time searching for the notes you need if they are not stored properly.
- Folders can be a useful way of storing all your notes on a particular topic – either hard copies in a plastic or cardboard folder or as a file digitally in a folder on the computer.
- Ring-binders can be a useful way of storing your notes along with any other papers on a topic, such as handouts and worksheets. The advantage of a ring-binder is that you can keep the notes together more easily than in a plastic or cardboard folder.
- Cards are a useful way of storing key information that can be used when reporting your findings or quickly referred to when revising for a test. The app Flashcards can be introduced to show the children how they can make and store information on an iPad. The app Simple Note is also a useful tool that can be used to teach children how to make and organise notes on a tablet.

COMPARING NOTES

Put the children in pairs, so that they have a note-making buddy. Ask them to compare the notes they made while researching the same topic. What can they learn from comparing them? Whose notes are more organised? Does one of them contain more of the important information than the other?

Make copies of Jason's and Hanif's notes on conductors and insulators. Distribute them to groups and ask the group to discuss the differences between them and to decide who they think is the more skilled note-maker and why.

Conductors and Insulators

Only certain substances allow electric currents to flow through them. These substances are <u>conductors</u>. Metals are good conductors. Wires in modern buildings are made of <u>copper</u>.

<u>Pylons</u> carry very powerful electric currents. Wires on pylons are made of aluminium. Touching bare electric wires can kill you. Wires at home are covered in <u>plastic</u>, which electric currents can't go through. Plastic is an <u>insulator</u>; so are rubber, wood, glass and stone.

Figure 4.3 Jason's notes

Conductors

Electricity can pass thru conductors e.g. metals – copper.

Pylons wires w. v. powerful currents – aluminium – can cause death.

Insulators *Insulators stop current passing thru e.g plastic (used to cover wires at home), rubber, wood, glass and stone.*

Figure 4.4 Hanif's notes

ASSESSING YOUR NOTE-MAKING SKILLS

Use this self-assessment sheet to find out how good your note-making skills are. Put a tick beside a statement if it is something you do regularly, a cross beside it if it isn't something you do and a ? if you do it sometimes.

1. I use my own words when I am making notes.
2. I use a highlighter or underline keywords and phrases.
3. I use headings and sub-headings.
4. I use abbreviations and symbols.
5. I avoid copying out whole sentences.
6. I make sure my notes are in a proper sequence.
7. I draw diagrams if appropriate.
8. I review the notes I make and, if necessary, reduce them so that I can revise them more easily,
9. I store the notes in a folder, a ring-binder, a card box or a tablet.
10. I keep the notes for different subjects separately.

What you learn from the assessment about your note-making skills:

Mostly ticks – You are developing good note-making skills.

A mixture of ticks and question marks – You have some good skills, but others need working on.

A mixture of question marks and crosses – Your note making skills are weak. You need to concentrate on improving your skills.

Mostly crosses – Your note-making skills are very poor.

CHAPTER 5
Developing your speaking and listening skills

Explain that in order to learn effectively you need to develop your oral skills as well as your reading and writing skills.

- You need to be able to join in group discussions which enable you to clarify your thoughts by providing you with the opportunity to explore ideas.
- You need to develop your skills as a speaker, so that you can express your ideas and report what you have found out in your investigations.
- You need to develop your listening skills so that you can become an active and attentive listener.

RULES FOR TALKING IN GROUPS

Explain that for us to be able to learn from each other by talking in groups we need to agree on rules that the members of the group will follow. Invite groups to draw up a set of rules that they agree upon. Prompt them as necessary to think about the importance of focusing on the question or topic, taking turns, not interrupting, respecting each other's point of view, agreeing to disagree, allowing everyone to join in and not sabotaging the discussion by talking about other things. Encourage the class to share the lists of rules drawn up by each group, then to agree on a list of rules for group discussions and to put it on the class noticeboard.

TALKING POINTS ABOUT GROUP DISCUSSIONS

Ask groups to discuss each of these talking points in turn, saying why they agree or disagree with them:

- Group discussions give you an opportunity to express your views.
- You can learn a lot by listening to other people's arguments.
- It is important to join in group discussions.
- You should respect other people's views, however strongly you feel about what they say.
- You should back up your views with reasons or evidence.
- You should be prepared to change your mind during the discussion.
- Group discussions tend to be dominated by one or two people.
- You shouldn't interrupt when someone else is talking.
- Group discussions give you a chance to explore ideas.
- It is annoying when people in the group don't stick to the subject being discussed.
- I'd much rather read about a subject than have to listen to people talking about it.
- In group discussions you get to hear things you hadn't thought of.
- Group discussions make you think.

TALKING IN GROUPS

There are opportunities for talking in groups in all areas of the curriculum. You can encourage the children to develop their skills by asking them to do the following activities:

In literacy
To discuss their responses to a poem, its meaning and the words the poet has chosen and what effect they have.

To predict what will happen in a story.

To add to a story by suggesting an alternative ending.

To choose a scene from a novel and act it.

In PSHE
To role play a studio discussion of a controversial topic.

To prepare for a debate.

In numeracy
To make a set of probability cards for the children to put into four groups: impossible (e.g. you will grow a beard overnight); unlikely (e.g. it will snow in June); likely (e.g. there will be dodgem cars at the fair); certain (e.g. you will have a half-term holiday).

To compare different containers according to their shapes and sizes.

In religious education
To compare creation stories in different religions and to discuss similarities and differences.

To investigate different religious ceremonies.

In music
To select appropriate music to introduce a radio programme about a theme such as Hallowe'en.

To compose a piece of music in a particular form, e.g. a rap.

In art
To discuss different artistic styles, such as impressionism and cubism.

To investigate a particular artist, such as Van Gogh, Picasso, Monet or Salvador Dali, and to prepare a PowerPoint presentation about him.

To choose artwork to put on display in the school entrance hall.

In history
To research and discuss what life was like in the trenches during World War I or for Londoners during the Blitz.

To plan and make a model of a Norman castle.

In geography
To discuss environmental issues, such as the pros and cons of the development of wind farms.

To plan a trip to a European capital.

SUMMARISING TALKS

Explain that these activities are designed to develop your skills as a listener by giving a summary of a talk.

Invite an adult to give a talk to the class. It could be about the topic you are studying, or a parent or grandparent talking about a subject in which they have an interest in or a person talking about their job. Tell the children that during the talk you want them to listen attentively but not to take notes. After the talk you are going to ask them in groups to summarise the main points of the talk, noting down what they can recall. They can then compare their notes to see which group has picked out the main points most successfully.

This is an activity which you can also use following an assembly, at which there is a speaker.

ACTIVE LISTENING

Ask the children individually to complete this self-assessment questionnaire on active listening by answering the questions by putting a tick in either the usually or sometimes column. Then discuss with a partner what this tells them about how active a listener they are.

	Usually	Sometimes
Do you concentrate on what other people are saying?		
Do you adopt a listening position by sitting up straight and not slouching?		
Do you look directly at the person speaking and maintain eye-contact?		
Do you show that you are listening by your facial expressions, e.g. a nod, a smile or a frown?		
Do you make a mental note of the main points others make?		
Do you give others feedback by summarising what they have said?		
Do you seek clarification by asking questions?		
Do you let others finish speaking and wait until it's your turn?		
Do you avoid shouting and interrupting?		
Do you refrain from fidgeting, doodling, shifting about in your seat or looking round the room or out of the window?		
When it is your turn to speak, do you refer to what other speakers have said?		

CLARIFYING

This is a simple way of fostering active listening when you are giving explanations or instructions to the class or conveying information, e.g. by showing a video. It involves pausing the video or pausing during your explanation and asking the children in pairs to explain to each other the key point or key points that have just been made.

Play the class a recording of one of the BBC's hourly news bulletins. Encourage the children to listen attentively without taking notes and then ask them to work in pairs. One of them repeats as much of the news broadcast as they can to their partner. When they have finished (but not before) the partner adds anything that he thinks they should have mentioned. Then play another similar recording and repeat the activity with the other person trying to recall as much of the broadcast as they can. You can do similar activities by asking them to listen to passages you read to them.

SPOT THE DIFFERENCE

Divide the class into two teams. Explain that you are going to read a paragraph from a story. Then the teaching assistant is going to read the same paragraph but she is going to change one detail. The children have to spot the difference by putting their hand up. The team to which the person who spots the difference first belongs is awarded a point. If a person puts their hand up, but has failed to spot the difference correctly, then the team to which they belong loses a point. You continue reading several paragraphs. The team with the most points is the winner.

THE SCHOOL BUS

Make sure everyone in the class has a piece of paper and explain that you want them to take notes on everything you say.

Explain that they are the driver of the school bus. Go on to tell them the number of children who get on at each stop and their ages. For example, at stop 1, three children get on; one aged 8 and two aged 10. At stop 2, five children get on; one is aged 7, two are aged 9 and two are aged 11. At stop 3, four children get on; two are aged 8, one is 10 and one is 11.

Then ask the children how many children are on the bus and how old the driver is. The answer is 13 – the 12 passengers, plus the driver who is their age.

TALKING POINTS ABOUT LISTENING

Ask groups to discuss each of these talking points in turn and to say why they agree or disagree with them:

Hearing is the same as listening.

Listening skills are hard to learn.

A good reader is always good at listening.

Young people are not as good listeners as older people.

You can learn how to become a good listener.

A good listener will learn more than a poor listener.

Good listeners are less likely to be prejudiced than poor listeners.

An active listener thinks about what people say.

Learning to listen is as important as learning to read.

You spend more time listening than you do reading, writing and speaking.

To be a good communicator you need to be a good listener.

HOW NOT TO LISTEN

This is an activity for pairs. Ask the children to decide which of them is A and which is B. Separate the As from the Bs. For example, take all the Bs out to the playground while the teaching assistant looks after the As. Ask the teaching assistant to tell the As that when the Bs come back they have to talk to them about what their favourite TV programmes, films, books and computer games are, and why they are their favourites. Their task is to try to persuade the Bs that they really should watch the programmes and films, and read the books. Meanwhile, you tell the Bs that they are to behave as though they are not interested in what their partner is telling them by yawning, looking out of the window, fidgeting and not paying any attention to what they are being told. After they have role played the situation discuss how the As felt about being ignored and what they learned from this activity about the way someone acts when they are not listening.

FOLLOWING INSTRUCTIONS

Shapes and colours

For this activity, each child needs to have five different coloured pencils (red, green, blue, orange, black) and a blank piece of paper on the table in front of them. Explain that they are not allowed to pick up any of the pencils until you tell them to do so and that you are going to ask them to draw some shapes and that you are only going to tell them what to do once. This is a test so they must remain silent. Then read this instruction. 'Please draw me a red circle, a blue square, an orange triangle, a black tick and a green cross. You may pick up your pencils now.' When they have finished, discuss how easy or difficult they found the task. Count how many children drew all five shapes correctly.

You can repeat the task giving instructions in which they have to use different colours for different shapes, e.g. a blue cup, a red plate, a black bowl, a green spoon and an orange fork.

Draw the picture

This is a barrier game for pairs. Give one of them a picture, making sure that the other person does not see it. The person with the picture has to describe it, so that the person listening can draw it. You can allow questions if you wish, but the importance of listening carefully is increased if you do not allow them to ask questions.

PLANNING AND DELIVERING ORAL PRESENTATIONS

Explain that often you are asked to give an oral report saying what you learned during a project.

Use this assessment sheet to think about how well an oral report was planned.

- Did it have a clear beginning?
- Did it have a beginning, such as a question, which captured the listeners' attention?
- Was it divided into separate sections?
- Did the speaker give examples to back up the points in the talk?
- Did the report end abruptly?
- Was there a firmly stated conclusion?
- What targets for the future could the speaker set?

Use this checklist to assess how well a talk was delivered.

- Did the speaker stand up straight?
- Did the speaker maintain eye-contact with the audience?
- Was the pace of the talk too fast or too slow?
- Did the speaker hold the audience's attention?
- Did the speaker use cue cards?
- Did the speaker mumble?
- Could the speaker be heard? Did they speak loudly enough?
- Did the speaker say 'um' or 'er' too often?
- Did the report end with a conclusion or did it just peter out?

HOTSEATING AND ROLE PLAYING

Encourage the children, if appropriate, to think of how they might express what they have discovered during a project using role play. For example, they might role play a public meeting to protest about a local planning issue or if a general election is imminent role play candidates from the main political parties and hold a mock election. They could hotseat a character from a book they are studying or interview an eye-witness of an event, such as the Battle of Hastings. If you are studying the Great Fire of London you could use the app Tellegami to help the children prepare an interview with an eye-witness.

CHAPTER 6
Developing computer skills

RESEARCHING ON THE INTERNET

Start with a group discussion of how the children use the internet. Ask: Do you find it easy to extract information from the internet? Do you get frustrated because often you cannot find what you want? What advice would you give to someone of your age as how best to find information on the web? What would be your top tip?

USING A SEARCH ENGINE

Explain that to find out information from the internet you have to use a search engine, such as Google. You need to think carefully about the keywords that you enter. The keywords need to be as precise as possible, otherwise your search will produce irrelevant information.

For example, if you are researching volcanoes and want to find out which ones have erupted recently, you will need to type in the keywords 'volcanoes', 'active', 'recent' and 'eruptions'. If you type in only 'volcanoes' and 'eruptions' you would get more results than you need and many of them would not be relevant.

It is often worth typing in 'kids' as well as your keywords. This can help you to find websites that are designed to provide material that is geared towards children and you can, therefore, avoid visiting sites that are aimed at adult researchers.

CHOOSING WHICH LINKS TO CLICK ON

Explain that if you have chosen your keywords carefully, the top links are the ones that are most likely to contain the information you are looking for. However, there are often adverts listed before the web results, which it is best to ignore. And it is worth scanning down the list of web results, since the ones you need may not be at the very top.

Before clicking on a link, look at the actual URL address. Ask yourself whether it sounds reliable. Is it a well-known site? For example, the National Geographical website would be a reliable source of information on volcanoes, whereas a site named volcanictruth, which claims to give 'The real reasons for volcanic eruptions' is likely to be unreliable.

DON'T BELIEVE EVERYTHING THAT YOU READ

Explain that you need to be critical of any information that you get from the internet. Ask yourself whether what you are reading is fact or opinion. Check the domain name of any site that you visit. Is it a site that you can trust? Does the organisation that runs the site have its own agenda?

If the address ends with .sch.uk, .ac.uk or gov.uk then it is an educational or government website and so it is one you can trust.

Anyone can buy a domain name and those that end with .co.uk, .com or .org have been bought. This doesn't necessarily mean that you can't trust them but some of them may contain biased information. Point out that you should look for sites belonging to organisations you can trust, such as the BBC, Parliament or the Natural History Museum.

Check the date the information was posted and whether or not the site is regularly updated.

If you are unsure whether or not the information you have obtained from a site is accurate, double-check it by visiting another site.

INFORMATION FROM APPS

Explain how information can be obtained on many topics from apps that are available for iPads.

For example, the app The Human Body by Tinybop has information on all the key parts of the body and their functions. There are many apps that can be used with history topics such as Timeline WW2 with Dan Snow and Pyramids 3D and the app Expeditions which can be used to explore not only places such as deserts and countries but also under the sea.

AN INTERNET SEARCH

Encourage the children to do an internet search on a topic such as the ancient Olympic Games or Egyptian writing and to report to the rest of the class which sites they found the most useful. Then hold a discussion about how they found these sites. Did they come across them by chance or did they find them because they searched systematically? Ask: Did you visit any sites that were unhelpful? Were the most useful sites ones that you could trust?

HOW TO USE THE INTERNET

Encourage the children to produce a poster offering advice on how to use the internet to find information. You could list Ten Top Tips for how to use the internet or a list of Do's and Don'ts when searching on the internet.

Alternatively, they could produce a PowerPoint presentation, a leaflet giving detailed advice or a video offering advice. Or they could role-play a sketch or produce a comic strip in which Willie

Wizkid, a computer whizz-kid, offers advice to Frustrated Frankie who cannot find the information he wants.

MAKING A FILE

Explain that once you have found the information that you have been searching for, you need to store it on the computer in a place where you can find it. You can store it by making a file on your computer. To do this:

- Open Microsoft Word.
- Click on the File button.
- Click on New.
- Click on Blank document.
- Type in any information you want to save.
- Click on the file button and click on Save As.
- Click Browse.
- Give your document a name, e.g. Volcanoes.
- Click on Save.

Explain that you now have a file labelled Volcanoes saved in the documents on your computer. You can type anything you find out about volcanoes into this file.

Alternatively you can select information you find on a website to store on the file by copying and pasting it into your file. Select the information you want to store by highlighting it, then press ctrl + c. When you go to the file where you want to store it, press ctrl + v and the passage will appear on the screen and you can save it in the file. This can be a useful way of storing information for reference, but remind the children that whenever they are asked to write about a topic they should always use their own words.

MAKING A FOLDER

Making folders on your computer is very important as it keeps your files tidy so you don't lose them. To make a folder is very simple:

- Use the mouse to point to an empty area on the desktop.
- Right click and then click on New.
- Click on Folder.
- Give it an appropriate name (for example, school).
- Press the enter key.

Encourage the children to use the internet to research a topic, such as the Romans in Britain and to build up a computer folder consisting of notes on various topics such as the reasons for the invasion, Hadrian's wall and Roman villas. Below are suggestions for the questions they could seek to answer.

The Roman Invasion
Links:

www.bbc.co.uk/schools/primaryhistory/romans/invasion/

1   When did the Romans invade Britain?
2   Why did the Romans come to Britain?
3   Why did the Romans defeat the Celts?

The Roman Army
Links:

www.primaryhomeworkhelp.co.uk/romans/soldiers.html

www.ducksters.com/history/ancient_rome_army_legions.php

4   How was the Roman army organised?
5   What did a Roman soldier wear?
6   What weapons did a Roman soldier carry?

Hadrian's Wall

Links:

http://primaryfacts.com/1540/facts-about-hadrians-wall/

www.historylearningsite.co.uk/ancient-rome/hadrians-wall/

7   Where did the Romans build Hadrian's Wall?
8   When was it built?
9   Why did they build it?
10   What was the wall built of?
11   What was a Roman fort like?

Roman towns

Link:

www.primaryhomeworkhelp.co.uk/romans/towns.htm

12   What were Roman towns like?
13   What was an amphitheatre?
14   What was the forum?
15   What was the temple?
16   What were the baths like?
17   Name two towns which have Roman names.

Roman villas

Link:

http://history.parkfieldict.co.uk/romans/roman-villas

18   What was a Roman villa?
19   What was it like inside a Roman villa?

The Roman occupation of Britain

Link:

www.mylearning.org/roman-grantham/p-4459/

20   How long did the Roman occupation of Britain last?
21   When and why did it come to an end?

FURTHER RESEARCH ON THE ROMANS

Talk about how we know a great deal about Roman homes and how the Romans lived from the remains of the town of Pompeii that was buried by lava when Mount Vesuvius, a volcano in south Italy, erupted. Encourage them to use the internet to find sites about Pompeii. Then compare their list of sites with those found by other members of the class and produce a PowerPoint presentation giving information about how you can use the internet to find out about Pompeii.

There are apps available that can help the children to learn about the Romans and how they lived such as Virtual History Roma and Streetmuseum Roman London. Google Play has an app on the Roman Forum and the Kids Discover website has apps on Pompeii and the Roman Empire.

WORD PROCESSING SKILLS

Explain that in addition to using a computer for research, you can also use it to present your findings, e.g. by producing a printed document or a PowerPoint presentation.

Show the children how to use the capabilities of a word processor to alter the appearance of the text in a document, e.g. by choosing a different font. Explain that the font is the sort of type you use when printing out a document. You can change how each character, letter, number or symbol looks by altering the font.

Explain that you can also emphasise certain words by printing them in bold or in italic, by using a different colour or by underlining them.

PICTURES AND DIAGRAMS

Explain that they can also use the computer to find pictures and diagrams to include in their notes or PowerPoint presentations. For example, Nerys wanted to include pictures in her presentation on volcanoes. So she searched Google by typing in 'volcanoes' and found that it suggested she could do a search of 'volcanoes for kids' so she typed in 'for kids'. She found lots of images of volcanoes erupting so she chose one and a diagram of the various parts of a volcano.

So she copied them, and then pasted the photo and the diagram into the document she was preparing for her presentation.

Explain how you can use an iPad to prepare a PowerPoint presentation to report your findings at the end of a project. There are apps such as Prezi which you can show the children how to use to include images, commentary and text.

WORD PROCESSING TIPS

In pairs, prepare a PowerPoint presentation giving instructions on how to make the best use of the word processor. Include advice on how to choose fonts, how to highlight text, how to incorporate pictures and diagrams into a document, how to move blocks of text and how to position illustrations.

ASSESSING A POWER POINT PRESENTATION

Content

- Did the introduction make clear what the presentation was about?
- Were the slides presented in a logical sequence?
- Were all the slides relevant and interesting?
- Was the information easy to follow?
- Were a variety of fonts used?
- Were all the fonts readable?
- Was good use made of pictures?
- Were any diagrams, graphs or tables included? Were they readable?
- Was any of the content confusing?

Delivery

- Was the information presented enthusiastically?
- Did the presenter make good use of the slides, e.g. by using a pointer to draw attention to particular points?
- Did the presenter use any progressive disclosure techniques so that information appeared on the screen point by point?
- Did the presenter do anything that distracted the audience, e.g. fidget, pause because they got their notes muddled?
- Did the presenter stand up straight and appear confident?
- Did the presenter engage the audience and maintain eye-contact with them?
- Did the presenter vary the tone of their voice?
- Could the presenter be heard throughout the room?
- Did the presenter say 'um' or 'er' too often?
- Was the delivery either too fast or too slow?

CHAPTER 7
Developing writing skills

This chapter focuses on developing a step-by-step approach to writing reports on a topic they have been investigating and on how to write instructions, arguments and reports of science experiments. It contains activities to develop children's understanding of the features of different kinds of writing and stresses the importance of choosing the right register and writing in standard English, avoiding using slang and txtspk. Further activities include hints on how to improve spelling and punctuation.

THE WRITING PROCESS

Explain that whatever kind of writing you are asked to do, the stages of the writing process are similar.

Put the flow-chart Figure 7.1 on the board and discuss it with the class.

Points to make during the discussion:

- Your plan should always be flexible. Be prepared to alter your plan if you think of something important as you are writing.
- Don't try to include everything that you thought of while collecting your ideas. Be selective and if necessary leave ideas out if they don't fit.
- Reviewing means checking to see if you have left out any important points and making slight alterations to what you have written.
- Redrafting means altering the content and structure of your writing.

Stage 1 Collecting and selecting ideas. Making a draft plan

Stage 2 Writing the first draft

Stage 3 Reviewing and redrafting

Stage 4 Proofreading

Stage 5 Producing the final copy

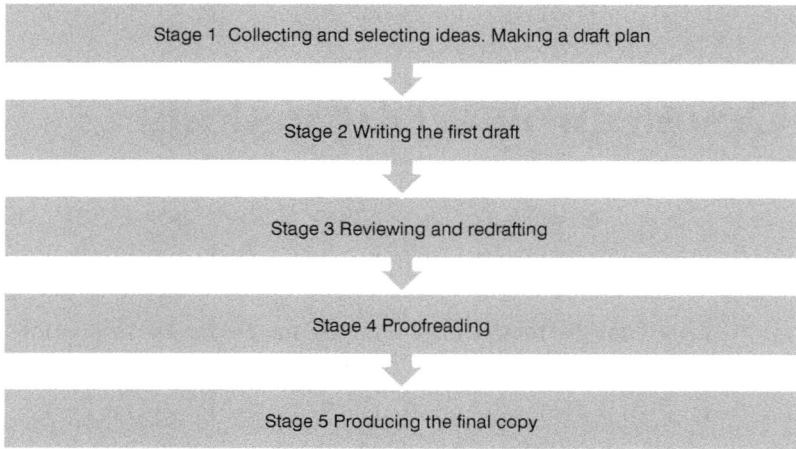

Figure 7.1 The writing process

- Proofreading means checking for errors of spelling, punctuation and grammar. It should not be confused with redrafting.
- The final copy should be a neat copy, which can be either handwritten or typed.

REPORT WRITING – A STEP-BY-STEP APPROACH

Discuss how many people find writing difficult because they do not adopt a structured approach to their writing. Talk about how a step-by-step approach to writing can help them to improve and develop their writing.

Put the five steps on the board and discuss each step in detail.

STEP 1 – Define the purpose and audience

- What is the purpose of your writing? Are you trying to tell a story? To describe a personal experience? To report your observations? To explain a process? To convey factual information or instructions on a particular subject? To express your opinion? To develop an argument?
- Who is going to read what you write?

Explain how the tone and style of your writing will differ according to your audience. If you are writing an e-mail to a friend you are likely to use a more informal style than if you are writing a report on an investigation to be read by your teacher.

Put these examples on the board, and discuss the difference in tone.

We went pond-dipping today. It was amazing. You'll never guess what we saw. There was a heron.

Birds that we observed included six ducks, two moorhens and a heron.

STEP 2 – Collect information and ideas

Explain that once you have defined the purpose and audience for your writing, the next step is to think about what information you might include. Look at any notes you have already made on the subject and do a brainstorm of all the ideas and information that you might include. Put down absolutely everything that you might include. It's better at this stage to have too much rather than too little. Don't worry about putting it in any particular order – you can do that at the planning stage (see Step 3).

Brainstorming information and ideas

Encourage the children to brainstorm all the ideas and information they could include when writing a report of an investigation they have been doing or a topic they have been studying.

STEP 3 – Planning

Explain that once you have brainstormed the information your written report might contain, you need to work out a plan which puts the ideas in the most logical order. Some people like to do this by writing numbers against the ideas they have brainstormed. Others prefer to make their plan as a flow-chart.

In addition to deciding on an order for the main points, you need to think of a good way of starting (an introduction) and a good way of finishing (a conclusion).

Making a plan

Encourage the children to make a plan in the form of a flow-chart, giving a paragraph by paragraph outline of what their written report on an investigation will contain.

STEP 4 – Writing a draft

Discuss how once they have made a plan, they can write a draft of your report. Explain that it does not matter at this stage if the ideas don't come out in the right order. Be prepared to alter your plan if something important occurs to you as you are writing.

Drafting a report

Encourage the children to write a draft of their report.

STEP 5 – Revising and redrafting

Explain that at this stage you are more concerned with the content, structure and organisation of their writing than with whether they have punctuated their writing properly or spelled the words correctly.

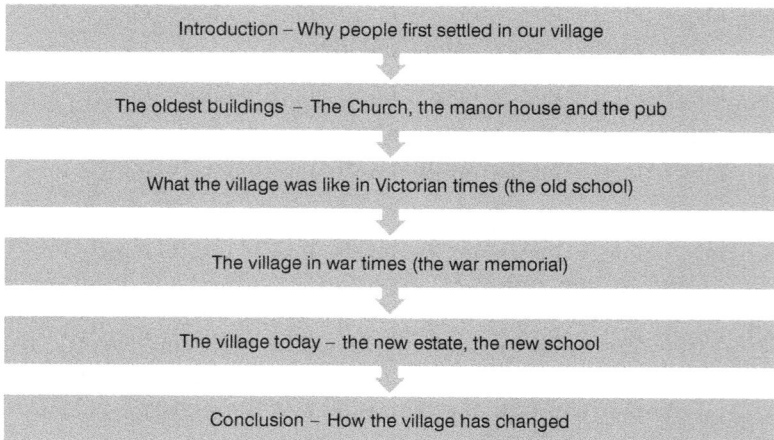

Introduction – Why people first settled in our village

The oldest buildings – The Church, the manor house and the pub

What the village was like in Victorian times (the old school)

The village in war times (the war memorial)

The village today – the new estate, the new school

Conclusion – How the village has changed

Figure 7.2 Example of a flow-chart plan: a report on the history of our village

Redrafting checklist
Encourage the children in pairs to read each other's drafts and to use this checklist to help them to suggest any changes they think their partner needs to make to their draft.

- Have they omitted any key points? If they have missed out something important, where should it go in?
- Are the points in the right order? Does it flow logically from one point to the next?
- Do they express themselves clearly? Are there any sentences that they need to revise because it is not clear what the writer is trying to say?
- Have they quoted any evidence to support their views?
- Is the introduction clear? Is it likely to catch the reader's attention? Discuss how it could be improved?
- How does the report end? Is the conclusion clear? Are there any questions that are left unanswered?
- Is the style of the report appropriate? Has the writer used standard English and avoided using slang?
- Overall, is the report presented well?

STEP 6 – Proofreading

Discuss how before they hand in their report, they must check the spellings and the punctuation.

STEP 7 – The final copy

Produce a neat copy either handwritten or on the computer.

DRAWING PIE CHARTS

Explain that presenting information visually as a pie chart can often be a useful way of reporting some of your findings. Remind the children what a pie chart is (see Chapter 3, p. 33).

OCEAN POLLUTANTS

Provide them with the information (below) which details the different amounts of the main pollutants entering the oceans. Encourage them to present the information in a pie chart. Explain that they will need to use the percentages to calculate the angle of each sector of the chart.

Percentage of different pollutants entering the oceans:

Sewage 30%; air pollutants 20%; farm run-off 20%; industrial waste water 10%; marine transportation 10%; offshore oil 5%; litter 5%.

ASSESSING A REPORT

Explain that this checklist can be used to assess a non-chronological report.

- Does the report have a title?
- Has the writer used headings and sub-headings?
- Is there a clear opening sentence stating what the report is about?
- Are there separate paragraphs focusing on different parts of the report?
- Is the style of the report formal and impersonal?
- Does the report contain facts rather than opinions?
- Does the writer quote evidence or give examples to support statements?
- Are technical terms included and explained?
- Does the report include any pictures, diagrams or graphs?
- Are the diagrams and graphs clearly labelled? Are there captions to any photographs?
- Is there a concluding statement?

CARBON FOOTPRINTS

Explain that a person's carbon footprint is a calculation of how much CO_2 they produce each year and, therefore, how much they contribute to global warming. It is measured by each tonne of the gas we produce.

Provide the children with the information (below) which details the different amounts that various activities contribute towards a person's carbon footprint Ask the children to present the information as a pie chart.

Home – gas, oil and coal 15%; Home – electricity 12%; private transport 12%; holiday flights 6%; food and drink 5%; clothes and personal effects 4%; car manufacture and delivery 7%; house buildings and furnishings 9%; recreation and leisure 14%; financial services 3%; share of public services 12%; public transport 3%.

DRAWING BAR CHARTS

Explain that a bar chart can be a useful way of reporting your findings. Remind the children what a bar chart is (see Chapter 3, p. 34) and that a bar chart:

- must have a title;
- has a horizontal line, known as the x-axis, and a vertical line, known as the y-axis, both of which must be labelled;
- must have a scale on the y-axis;
- is easier to read if there is a gap left between each of the bars.

WHERE WE WERE BORN

Ask each of the children in a class to write down the name of the country in which they were born on a slip of paper. Make a pile of the slips for each different country and then count the numbers in each pile and write the names of each country and the number of children born there on the board. Then ask the children to present the information as a bar chart.

HOW WE GET TO SCHOOL

Ask the children to write on a slip of paper how they travel to school – by car, by bus, by minibus, by taxi, by train, by bike, by walking. Collect in the slips, count them and put the numbers for each method on the board. Invite the children to present the information as a bar chart.

A 100-WORD REPORT

At the end of a project, ask the children to give a written report of the project in the form of a newspaper report using a maximum of 100 words. Remind them that a newspaper report is designed to convey essential information that answers the questions: Who? What? Where? When? Why?

WRITING INSTRUCTIONS

A set of instructions should always follow the same structure:

Title – e.g. How to make an African mask

Materials and equipment

Numbered steps + diagrams

As part of a design and technology project ask the children to write a set of instructions explaining how to make an object.

ASSESSING A SET OF WRITTEN INSTRUCTIONS

Use this checklist to assess a set of instructions that you or a partner has written.

- Is there a title which clearly states what you are going to make?
- Is there a list of the materials needed?
- Does the list include the quantity of each material?
- Is there a list of any tools and equipment you need?
- Are there step-by-step instructions explaining what you need to do?
- Are the steps numbered?
- Are there diagrams or drawings which make it easier to understand what needs to be done?
- Are the instructions written in the present tense?
- Are the sentences short and clear?
- Are there time connectives to show the order in which the steps are to be done, e.g. first, next, then?
- Have you used sentences beginning with 'bossy verbs'?

WRITING UP A SCIENCE EXPERIMENT

Discuss how the report of a science experiment has several sections. Go through the sections, explaining what each section must contain, then encourage the children to write a report on an experiment they have been doing.

1 **Purpose**
  This is a brief explanation of why you are doing the experiment.
2 **Hypothesis**
  This is a description of what you think will happen. You should include what you thought would happen, even if it didn't happen.

3 **Materials and equipment**
   List all the materials and equipment you need for the
   experiment. If appropriate, draw a labelled diagram.
4 **Procedure**
   Write a detailed description of what you did, giving a step-
   by-step account of the method you used.
5 **Record of observations**
   Describe what you observed, collecting and recording the
   data, presenting it in writing, drawing pictures, tables and
   graphs as appropriate.
6 **Analysis**
   Analyse the data to see what it tells you.
7 **Conclusion**
   Decide what you have found out from the experiment. Write
   a paragraph saying whether the data that you collected
   proved or disproved your hypothesis. Include an explanation
   of anything that you discovered.

WRITING AN ARGUMENT

Explain that a written argument not only states your point of view,
but should also include facts to support it and states why you
disagree with people who hold a different view.

Here is a typical plan for a written argument:

Paragraph 1: Introduction – Your point of view.

Paragraphs 2, 3, 4: Reasons for and evidence to support your view.

Paragraphs 5, 6: Different viewpoints and reasons why you
disagree with them.

Paragraph 7 Conclusion – The main reason why you hold your
view.

ASSESSING A WRITTEN REPORT OF A SCIENCE EXPERIMENT

Encourage the children to reflect on how successfully they have written about a scientific experiment by using this checklist. Ask them to look at a partner's work and to comment on what the person has done well, then to discuss with them and to decide what they need to do to improve their description of a scientific experiment.

Before carrying out the experiment:

- I clearly stated the aim of the experiment, saying what I wanted to find out.
- I listed the resources I would need.
- I explained my method, saying why I thought it was a fair way of seeing what happened.
- I explained how I would record my measurements and/or observations.
- I predicted what I thought would happen and why.

After carrying out the experiment:

- I explained the results of the experiment – either as a written account of my observations or in the form of graphs, charts or tables.
- I analysed the results, saying what I saw and noting any patterns.
- I stated what I found out.
- I stated whether the result was what I had predicted.

Explain that there are a number of techniques they can use when writing an argument to make it more effective. Put 'Tips for writing arguments' on the board and discuss each tip. Then ask the children to write an argument expressing their views about, for example, topics such as: we shouldn't have homework; should we keep animals in zoos?

TIPS FOR WRITING ARGUMENTS

Capture the reader's attention
Start with a statement or question that grabs the reader's attention.

For example, 'Children under 10 should not be allowed to cycle on their own.'

Quote facts and statistics
Quote facts and statistics to support your argument.

For example, 'Figures show the dangers of cycling for young people. In one year recently 85 children aged 8–11 were killed or seriously injured in accidents.'

Refer to personal experiences
This suggests that you really know what you are writing about.

For example, 'I've nearly been knocked over by people riding across the pavement without looking.'

Involve the reader
Involve the reader by addressing them directly.

For example, 'You need to check your brakes, lights and tyres regularly.'

Include questions
These can have a dramatic effect, especially if they do not require an answer.

For example 'Isn't it about time that we made cycle helmets compulsory?'

Undermine other arguments
Put the opposite point of view to yours and say why you don't agree with it.

For example 'Some people argue that cycling to school should be encouraged, because you get exercise. But there's so much traffic in the rush hour and not enough cycle lanes that it's just not safe.'

End with a strong statement
Make sure you end on a high note.

For example, 'Until we make it safer, children under 10 should not be allowed to ride on the roads.'

Vary your sentences
Explain that when writing arguments, you need to vary your sentences. Otherwise your writing may become repetitive and dull.

Put the following sentence openings on a chart for the children to refer to when they are writing or when playing the Sentence starter game.

Sentence openers

Also... Although... Because... But... Clearly... Consequently... Despite... Even though... For example... Furthermore...

However... If... In addition... In conclusion... In fact... In spite of... Maybe... Moreover... Nevertheless... Obviously... Perhaps... So... Sometimes... Therefore... While...

SENTENCE STARTERS

This is a game for any number of players. Player 1 starts the game by choosing one of the openings and saying a sentence which starts with that opening. For example, they could choose 'since' and begin the game by saying: 'Since it was raining, Jamal decided to stay indoors.' The second player then has to add a sentence starting with another sentence opener. For example, they could choose 'although' and say: 'Although there was nothing to do except watch TV.' The third player has to choose another sentence opener and the game continues until a player is unable to add a sentence starting with an opener that has not already been used. That person then drops out of the game which continues until there is only one person left, who is the winner.

SENTENCE STRUCTURES

Explain that different sentence structures are used in different types of writing. In this exercise the children have to match the sentence to the text type. Ask them to match the sentence to the text type.

Sentence	Text type
Put the stick through the hole in the cardboard and fix it with tape.	Narrative
Buy one book. Get another half price.	Information
Hedgehogs hibernate during the winter months.	Recount
Therefore, in my opinion boxing without headgear is too dangerous and should be banned.	Explanation
Yesterday, we went on a trip to the museum.	Persuasion
The stone sinks because it is too heavy.	Argument
Teresa turned round and gasped with horror as she saw a hand appear.	Instruction

STANDARD ENGLISH

Explain that, whatever kind of writing you are asked to do, it is important to make sure that you use standard English and that it is grammatically correct. What is acceptable in speech may not be acceptable in your written work.

Put Kenstone's recount of pond-dipping on the board. Explain what a dialect is – a type of English spoken by a particular group of people – and that because Kenstone speaks a dialect that is different from standard English, he uses that dialect when he writes.

Pond-dipping

We went pond-dipping. There weren't no newts. But we saw some of them tiny fish. Miss say they called sticklebacks. We catch some in a net, but Miss tell us to put them back. We was hoping to see some frogs but there weren't none.

— Kenstone

Ask the children in groups to rewrite Kenstone's recount in standard English.

Then discuss the changes that they made and point out that:

- All the verbs need to be in the past tense say/said, catch/caught, tell/told.
- There weren't no newts is a double negative and 'no' should be 'any'.
- 'Them' should be 'those' in the phrase 'them tiny fish'.
- 'We was hoping' should be 'we were hoping'.
- 'There weren't none' should be 'there weren't any'.

GRAMMAR CHECK

Explain that most computers have a grammar check. If you are writing on a word processor, a grammar check is usually switched on automatically and a green line appears underneath where the computer thinks you have made a mistake. You can then correct it and the green line disappears. However, if you cannot understand what the computer thinks you have done wrong, then right click on the mouse. The computer will then give you what it thinks you should have written. If you think the computer is right, you can then change what you have written. But remember the computer can make mistakes too and be prepared not to accept what it says if you think it is wrong.

SLANG AND TXTSPK

Explain that it is important whatever type of writing you are doing to avoid using slang. Discuss what slang is – a type of language used informally, e.g. 'copper' when referring to a police officer or 'quid' meaning a pound.

Put this extract from a report on the board and discuss how the use of slang is inappropriate.

We met this funny old chap who gave us a right dressing down and started to tell us to clear off. But when we told him about our project, he changed his tune and we were gobsmacked by what he told us. It turned out that he was a pal of Mr Stevens. What that old bloke knew was awesome!

Explain to the children that while it is perfectly appropriate to use abbreviations in notes (see pages 42–45), it is inappropriate to use txtspk in their written work, which must always be in standard English.

ASSESSING WRITING

The following activities can be used following any type of writing the children have been doing.

Ask the children to work in pairs, which you have carefully chosen, so that each person has a 'writing buddy'.

SELF ASSESSMENT

Invite the children to be involved in the assessment of their writing by writing a comment at the end of their work before handing it in. Encourage them to say which parts of the writing went well and which parts they think need to be improved.

PEER ASSESSMENT

Ask the children to hand their work to their writing buddy, who in turn will identify which parts of the writing they think work well by putting a tick beside them. They should put an asterisk beside any part they think needs improving and a comment suggesting how it might be improved. They should also put a question mark against any part of the writing they do not understand. Alternatively, they could use different coloured highlighters – green for the things that are done well, blue for the areas that need improvement and red for the parts they did not understand.

TEACHER ASSESSMENT

The next step in the assessment process is for you to write a personal response to each individual commenting on what they have done well and what they need to work on when they next do a similar piece of writing.

SETTING TARGETS

It is important for you to allow time for each individual to reflect on your comments and to set themselves targets for improving their writing. In addition to them setting targets for themselves, you could see them individually to discuss their targets, e.g. during a private reading session.

DEVELOPING SPELLING SKILLS

Highlight the hard part
Encourage the children to identify the part of a word which makes it difficult to spell. Here are some words with the difficult part highlighted:

Wednesday; encourage; Parliament; every; **phy**sical; app-**laud;**
muscle.

Split it into syllables
Explain that splitting difficult words into syllables can make it
easier to remember how to spell them. For example: at-ten-dance;
in-dust-ry; in–ter–est; re-mem-ber; ap-point-ment; part-ic-u-lar.

Mispronounce difficult words
Explain that deliberately mispronouncing a word can help you to
remember how to spell certain words. For example, **BUS-I**-ness;
cat-a-log-**U-E;** gar-**AGE**; lang-**U**-age; **KNOW**-ledge; **COL-O**-nel.

This can be particularly useful when words have a silent letter in
them e.g. thum**B**; **K**nuckle.

SIXTY SECOND SPELLING CHALLENGE

Split the class into groups of four or five. The aim of this game is
for the groups to think of as many words as they can which end
with the same string of letters within the time limit of 60 seconds.
Explain that during the 60 seconds there must be complete silence
while each member of the group writes down as many words as
they can think of which end with that string of letters. After the
60 seconds is up, the groups produce a group list of all the words
the people in the group thought of. Each word is worth one point,
but the group loses a point if they have included a word that isn't
spelled correctly.

Among the letter strings that can be used for this challenge are:

-ack -ate -end -ick -ide -ore -ump

-able -ation –ious -ight -ment -ough

THE SILENT LETTERS CHALLENGE

This is a group activity. Give each group a large sheet of paper and ask them to draw columns on it and to label the columns:

Silent B; Silent C; Silent G; Silent H; Silent K; Silent N; Silent T; Silent W.

Set a time limit and ask the children to list as many words as they can think of which include those silent letters in them. Point out that the silent letter may come either at the beginning of a word, e.g. wrist, at the end of a word, e.g. lamb or in the middle of a word, e.g. scent. Encourage them to use a dictionary to find words and to check spellings. The winning group is the one that has listed the most words.

When they have completed the challenge, get groups each to make a chart to put up for display. Note: Asking them to make their own charts, rather than putting up a printed copy of a chart, will reinforce how the words are spelled.

THE SOUND OF SILENCE

This is a fun activity which involves emphasising silent letters. Go round the class choosing words at random from your list of words with silent letters asking individuals to pronounce the words as they would be pronounced if the silent letter wasn't silent, e.g. knock would be ker-nock; castle would be cast-tel; lamb would be lam-bee.

SILENT LETTERS CROSSWORD

Explain that each of the words in the answers to this crossword contains a silent letter.

Clues across:

1. Your laces won't undo if they are tied in one (4)

2. The part of your body that joins your hand to your arm (5)

4. A medieval soldier (6)

6. What a workman who fits water pipes does (5)

7. What a dog will do to a bone (4)

Figure 7.3 A crossword

9. A large fortified medieval stone building (6)

11. What a person looks like when they are serious (6)

14. A small bird (4)

15. An explosive device (4)

Clues down:

1. To make something out of wool (4)

2. To put paper round a parcel (4)

3. A place where someone is buried (4)

4. A finger joint (7)

5. Trustworthy (6)

8. A joint in your leg (4)

9. A small piece of bread (5)

10. What you should do when someone is talking to you (6)

12. A young sheep (4)

13. When you have no feeling, e.g. when your hand is very cold (4)

Answers

Across: 1. knot; 2. wrist; 4. knight; 6. plumb; 7. gnaw; 9. castle; 11. solemn; 14. wren; 15 bomb.

Down: 2. wrap; 3. tomb; 4. knuckle; 5. honest; 8. knee; 9. crumb; 10. listen; 12. lamb; 13. numb.

ACROSTICS

Explain what an acrostic is – a number of lines or words in which the first letters in each line or word together form a word. Give an example of how an acrostic can be used to help you to remember how to spell a difficult word:

Big Eyes And Ugly Teeth I Find Utterly Lovely

Confusion Havoc Atrocity Outrage Shock

Foxes Often Roam Everywhere In Great Numbers

Invite groups to make up their own acrostics to remember difficult words such as: enthusiasm, league, separate, cocoa, yacht, reminisce, neighbour, ghost, hymn, restaurant, campaign, hieroglyphics, bureau, silhouette, biscuit, trousers, limousine, meringue, awkward, rhyme, rhythm, science, conscience, pioneer.

PUNCTUATION

Discuss why we use punctuation and the difficulties we may have when trying to read a piece of writing if it is not punctuated properly.

Hold a class discussion about when to use capital letters and full stops. Make copies of this passage and ask individuals to punctuate it properly.

it was 15th september 1830 a huge crowd of 50,000 had gathered to watch the opening of the liverpool and manchester railway it included the prime minister and the mp for liverpool called mr william huskisson they had come to see the engine built by george stephenson and his son robert it was called the rocket there was a terrible tragedy mr huskisson stepped onto the track to look at one of the engines the rocket came down the track where he was standing he was knocked over and killed

PUNCTUATION PUZZLES

This is an activity for pairs or groups. You need to prepare for this activity by making small squares of card each with a punctuation mark on it. Put a pile of the cards on each table – you will need lots of full stops and commas per table, but fewer semi-colons, colons, questions and exclamation marks. You also need to make copies of a paragraph without any punctuation marks, leaving a gap where a punctuation mark should go, making enough copies for each pair or group to have one. The children have to pick the appropriate punctuation marks from the piles and place them in the gaps. It is suggested that you use double spacing and leave plenty of space for the children to place the cards in.

PUNCTUATION ON PARADE

For this activity you need to make large cards each with a different punctuation mark on it. As well as a full stop, a comma, a semi-colon, a colon, a question mark and an exclamation mark, you could also make cards with an apostrophe and CAPITAL LETTER on them. Give out the cards to individuals to hold and ask them to stand in front of the class facing the board. Put a passage on the board and read it aloud to the class including the punctuation marks as they appear. Whenever the punctuation mark they are holding occurs, the individual holding that card must turn around and face the class.

Ask the children to work in pairs. Get them each to choose a paragraph from a non-fiction book and to copy it out leaving out all the full stops and capital letters. Encourage them to choose a paragraph that has the names of people or places in it. Then ask them to swap extracts and to see how successful they were in replacing the missing punctuation. Note: this activity works best if it is done on a word processor as changes to a text can easily be made.

THE PUNCTUATION BLUES

Encourage the class to learn this poem in order to reinforce their knowledge of the different punctuation marks and their functions.

The Punctuation Blues

I got up this morning, but I should have stayed in bed,
For punctuation marks were whizzing round my head.
I woke up this morning – I was so confused.
I think I had a dose of the punctuation blues.

There were apostrophes here. There were apostrophes there.
Apostrophes were scattered everywhere.
Some were trying to give a lesson
On how they can be used to show possession.
While another group were milling about
Saying use us when a letter's left out.

The capital letters were causing a fuss,
Shouting, 'You must remember how to use us
At the start of names like Imran and Paul,
And the names of places like Hadrian's Wall.
And don't forget us, whatever you do
You must use us to start your sentences too.'

There were groups of commas in front of a list
Saying we mustn't be left out, omitted or missed.
We'll show you in sentences where there are pauses
By indicating breaks between phrases and clauses.

A pair of exclamation marks were dancing about
And in a loud voice I heard one shout,
'Stop! Come here! Do as I say!
Don't do that! Get out of my way!'

There were question marks everywhere too,
Asking: Where? What? When? Why? Who?
There were semi-colons looking for clauses
Between which they could show longer pauses.

There were hyphens playing hide-and-seek
And speech marks waiting for people to speak.
While circling the keyboard awaiting commands
To surf the net were a thousand ampersands.

I got up this morning, but I should have stayed in bed,
For punctuation marks were whizzing round my head.
I woke up this morning – I was so confused.
I think I had a dose of the punctuation blues.

CHAPTER 8
Organisational skills – time management and teamwork

This chapter consists of activities that involve the children in thinking about their homework habits and how they manage their time. It also contains a number of activities and challenges that are designed to develop teamworking skills.

HOW GOOD ARE YOUR HOMEWORK HABITS?

Explain that it is important for them to plan how they spend their leisure time so that they develop good homework habits. Ask individuals to complete this quiz to find out about their homework habits.

YOU AND YOUR HOMEWORK – A TEST-YOURSELF QUIZ

1   When homework is set do you write down the details in your homework diary? a) Usually b) Sometimes c) Never
2   Do you check your bag at the end of each day to check that you have all you need for that evening's homework? a) Usually b) Sometimes c) Never
3   Do you work out a plan each evening of how you are going to spend your time? a) Usually b) Sometimes c) Never

4   Before starting your homework, do you check in your homework diary to see what the homework is and when it is due? a) Usually b) Sometimes c) Never

5   Do you find it hard to get started because you can't find the things you need? a) Usually b) Sometimes c) Never

6   Do you do your homework while watching TV? a) Usually b) Sometimes c) Never

7   Do you let yourself be interrupted when you are doing your homework? a) Usually b) Sometimes c) Never

8   Do you put off doing your homework until the last minute? a) Usually b) Sometimes c) Never

9   If you are stuck, do you give up rather than try to find someone to ask for help? a) Usually b) Sometimes c) Never

10   Do you read through your homework to check for mistakes before handing it in? a) Usually b) Sometimes c) Never

11   If you are learning something for a test, do you ask someone to help you by testing you? a) Usually b) Sometimes c) Never

12   Do you hand your homework in on time? a) Usually b) Sometimes c) Never

Make copies of the score sheet (below) and ask the children to work out their scores.

Add up the points for your answers from the list below:

1	a) 2	b) 1	c) 0
2	a) 2	b) 1	c) 0
3	a) 2	b) 1	c) 0
4	a) 2	b) 1	c) 0
5	a) 0	b) 1	c) 2
6	a) 0	b) 1	c) 2
7	a) 0	b) 1	c) 2
8	a) 0	b) 1	c) 2
9	a) 0	b) 1	c) 2
10	a) 2	b) 1	c) 0
11	a) 2	b) 1	c) 0
12	a) 2	b) 1	c) 0

Explain what the scores suggest about their homework habits.

19–24 suggests that they are developing good homework habits.

13–18 suggests that they are sometimes doing the right things, but they are not doing them regularly enough.

0–12 suggests that they either don't think homework is important or they haven't yet learned how to get themselves organised in order to get their homework done.

REVIEWING HOW YOU USE YOUR TIME

Encourage the children to carry out a review of how they spend their time after school in order to find out if they are using it efficiently. Make copies of the chart shown in Figure 8.1 and ask them to fill it in just before they go to bed showing how they spend the time each day between when they arrive home from school and the time they go to bed. Include activities such as playing computer games, texting/chatting on their mobile, using social media, watching TV and playing outside.

	Monday	Tuesday	Wednesday	Thursday	Friday
4.00–4.30					
4.30-5.00					
5.00–5.30					
5.30–6.00					
6.00–6.30					
6.30–7.00					
7.00–7.30					
7.30–8.00					
8.00–8.30					
8.30–9.00					

Figure 8.1 A weekday time chart

Discuss what they learn from the chart, identifying particularly any periods of time which they might have used more productively. Encourage them to think of time-wasting events such as:

- Replying to friends' messages when you are trying to study
- Spending too long playing computer games
- Too much time spent watching TV
- Playing football or time spent outside with friends
- Time spent finding out what homework you have
- Time spent searching for books and equipment you need.

Talk about how once they have identified any time-wasting events, they need to think of a plan of action that will enable them to eliminate them in the future. For example, they may have to limit the time they spend texting their friends, watching TV or playing computer games.

ORGANISE YOUR WORKSPACE

Discuss how in order to ensure that they get as much work as possible done in the time they have set aside for studying, they need to have a study area that is kept organised and free from distractions. Encourage the children to get in the habit of tidying away books and equipment at the end of each study session. This will enable them to stop the worktop from getting cluttered and enable them to find things quickly the next time they are needed.

AN IDEAL WORKSPACE

Invite groups to discuss where they do their studying. Ask: do you always do your studying in the same place or do you do it in different places? Do you have a desk or shelves on which you can keep your books and equipment? Do you have internet access from a tablet, iPhone, PC or laptop? Do you have to work in a room with a TV on and in which there are other people?

Set groups the task of designing what they consider to be an ideal workspace for someone of their age. Make a list of all its features.

Then ask them to compare the ideal study space with the place where they actually do their studying. Ask: is there anything you could do to make your study space more suitable to your needs?

GET YOURSELF ORGANISED!

Invite the children in pairs to role play a scene in which Organised Olly offers advice to Disorganised Dan on how to get himself more organised so that he makes better use of his time and not only hands his homework on time, but also achieves better grades. Encourage them to do the role play twice, taking it in turns to be the two different characters. As a follow-up activity, they can produce a poster offering advice about how to get organised in the form of a column of Dos and a column of Don'ts.

DEVELOPING TEAMWORKING SKILLS

Planning a playground

Put the children in groups. Explain that their task is to share ideas for the equipment they would put in a new playground in a park. They then have to design and make models of the equipment and prepare a presentation to give to a panel which is going to say how practical they think their suggestions are and to choose a winner. The panel could consist of, e.g. another teacher, a governor or a parent.

Planning a carnival float

Inform the class that they have been chosen to decorate a lorry as a float for the carnival which is taking place in your town or village. Tell the children that they can either decide on a theme themselves or choose one of the themes from this list: Favourite animals from books and films, The Four Seasons, Holidays, Our School, Myths of the Norsemen, Dinosaurs, Nursery Rhyme Characters.

Ask them how they are going to choose a theme. Once the theme is chosen, they need to share ideas on how to decorate the float. Who is going to design the decorations? How will they be made? What materials are they going to need? How will they get the materials? What jobs need to be done?

Emphasise the need for cooperation and that you need to have a plan in which you say who is going to do each job.

Once you have made your plan discuss how well the class worked as a team.

MASK-MAKING

Ask the children to work together to design and make masks. They could be Hallowe'en masks, animal masks, dragon masks or dinosaur masks. Encourage them to research masks on the internet to find different designs and to think about the questions they will need to answer in order to plan how to make the masks, such as: What materials will they need? How will the materials be joined together? How will the masks be attached to the face?

Ask the children to draw up a design brief and to share their briefs together, making suggestions, where appropriate as to how the brief might be improved.

Then encourage them to make the masks and to show them to the class, explaining what went well during the manufacturing of the masks and what they would do differently if they were to make masks again.

THE SPAGHETTI CHALLENGE

The Science Museum website has details of this activity in which the children have to work in teams of four or five to build a free-standing tower using only spaghetti and marshmallows that will hold a chocolate egg for 30 seconds without falling apart. The winner is the team that builds the tallest tower which successfully holds the egg.

In the debriefing session, the children can be asked to reflect on how successfully they worked as a team, considering such questions as: How well did you work together? Did you co-operate with each other or were there disagreements that held you back? Did you take on different roles? What would you do differently as a team if you were given a similar challenge in the future?

THE CATAPULT CHALLENGE

The website http://www.mylearning.org/crazy-catapults-inspired-by-the-romans-in-yorkshire/p-296/ has details of a challenge to design and build a Roman catapult.

MONSTER PHYSICS

Monster Physics is an app which can be used to develop teamwork, as it involves the children in solving problems such as building virtual machines for their monster.

THE ADVERTISING CHALLENGE

Ask the children to work as a team to design and make a box that is to contain LOCAL CHOCS a new brand of low calorie chocolates. Encourage them to think about such questions as what materials they will need, what shape the box will be and what writing will appear on the box. They can then make the box and present it to be judged. Invite another adult to role play a LOCAL CHOCS representative and to say whose box they would choose to use for their product and why.

A HALLOWE'EN DISCO

Invite the children in groups to draw up plans for a Hallowe'en disco. Questions that need to be answered are:

- Who is the party for? Will people from other year groups be able to attend?
- Where will the party be held? What do you need to do to book the venue?
- Will you need to have adult supervision? If so, who will provide it?
- Do you intend to decorate the room? If so, what decorations will you put up and who will provide them?
- Will you charge a small entry fee in case you have to pay for hiring a room?
- What refreshments will there be? Who will provide them?
- Will you have a fancy dress competition? If so, what prizes will there be? Who will provide them? Who will judge it?
- Who will be responsible for providing the music?

Encourage the children to discuss these questions, to share their ideas for the disco and how they would divide up the tasks that would need to be done prior to it, and then to choose one person to present the group's ideas to the class.

A BOOK WEEK CHALLENGE

Explain that the aim of the challenge is to create something that tells the story of a book, its themes and characters in a different form. It could be through drama, e.g. by creating a scene from the book in which the characters act out what happens or in which there is a different outcome; it could be through art, e.g. by making a collage or designing an advert; it could be by making a promotional video or radio advert in which you hot-seat characters and include extracts from the book; it could be by making model characters and making a scene to go in a cardboard box with other items that help to explain the story; or it could be by making glove puppets of the characters and re-enacting scenes in a puppet theatre. Explain that these are some of the ways that they could meet the challenge, but encourage them to come up with any other ideas of their own.

SHEPHERDS AND SHEEP

This is an activity to do in the hall. Prepare the hall by setting out some benches, chairs, tables and boxes randomly around the hall. Split the children into teams of equal size with between four or six in each group and ask them to sit on one of the mats you have placed round the edge of the hall.

Explain that all but one of the group is to be blindfolded and they are to kneel down and act as sheep. The person who is not blindfolded is a shepherd, who is unable to talk. The shepherd's job is to herd the sheep into a pen, which is one of the mats at the side of the room. He has to take the sheep through a field in which there are several obstacles such as mines, trip wires, barbed wire and electric fences, represented by the benches and chairs etc. which have been placed in the room. The group must decide how the shepherd is going to communicate with them, e.g. by stamping, clapping, whistling or tapping. Each group starts with 30 points and loses a point whenever one of them touches an object. If the shepherd speaks, the group loses 10 points. Allow them time to decide how they are going to communicate, before telling the shepherd which mat represents the sheep pen they must guide their sheep into. The first team to get all the sheep into the pen gets a 25 point bonus, the second team gets a 20 point bonus, the third team gets a 15 point bonus, the fourth team gets a 10 point bonus and the fifth team a 5 point bonus. The winning team is the team with the most points.

HOW GOOD A TEAM WORKER ARE YOU?

A test-yourself quiz

	Usually	Sometimes	Never
Do you remain silent in group discussions, letting others do the talking?			
Do you listen carefully to other people's opinions before making up your mind?			
Are you willing to compromise in order to reach an agreement?			
Are you willing to change your mind if someone suggests that you are wrong?			
Do you help the group to refocus if the group starts to talk about things that are nothing to do with the topic being discussed?			
If you put forward an idea and the rest of the group is against it, do you allow yourself to be outvoted?			
When no one else wants to take on a role, do you volunteer to take on the role yourself?			
If you agree to do something for the group, do you always do it?			
Do you interrupt when other people are talking?			

CHAPTER 9
Revising and preparing for tests

This chapter focuses on strategies that the children can use when preparing and revising for tests.

PREPARING FOR TESTS

Explain that before you start a revision session, it is important to decide not only what you are going to revise but how you are going to revise it. Your approach will depend on what you are going to be tested on and what form the test takes. If you are revising for a science test, you may want to find a test on living things. You can find tests on the internet at the BBC KS2 Bitesize Science website which has tests and notes on living things, including food chains, health and growth, plants and animals. If you are revising maths, identify which problems you are having trouble solving and concentrate on practising those problems.

Always plan your revision, so that it is most effective. Don't just sit down and start revising by reading and re-reading your notes over and over again in the hope that you'll gradually learn them. Not only is this boring, but it is not a good way of learning, because you are not actively involved in using the information that you are trying to learn.

A STEP-BY-STEP REVISION STRATEGY

Explain that this strategy can be used to revise any topic.

1   **Make a revision plan.** Decide which topics you are going to revise on which days. Stress that it is important to be realistic and to set achievable goals and not to plan sessions that are too long. It is better to plan short sessions with breaks, as you are less likely to remember information if you try to cram in too much at once and your mind becomes tired.

2   **Study your notes.** Read through the notes that you have on the topic, underlining or highlighting key facts and important terms.

3   **Make revision cards** consisting of the key information. Explain that condensing their notes helps them to think about and to understand what they are trying to learn.

4   **Learn it!** Study the revision card, using any methods of remembering things that work for you. (There is a list of memory tips opposite.)

5   **Test yourself** or get someone else to test you.

LEARNING STYLES

Explain that there are three basic learning styles – visual, auditory and kinaesthetic – and that using a multi-sensory approach assists children's learning.

MEMORY PLANS

Explain that there are various strategies we can adopt when revising and trying to memorise facts for tests. On your own, study this list of memory tips and pick out the tips you think are the most useful. In groups, discuss your choices and the reasons for them. Then each make a memory plan to help you to remember what you need to know for a test.

MEMORY TIPS

- Make a set of index cards or post-it notes with key words on them.
- Use different coloured pens or highlighters to identify key words and concepts.
- Review what you have learned one day, by looking at it again the next day. Frequently reviewing information helps to make it stick.
- Try learning the information by listening to it. For example, you may be able to find a video you can watch and listen to. Some people find it easier to remember what they hear rather than what they see.
- Repeat what you have learned by saying it aloud either to yourself or a revision buddy. Speaking facts aloud can help you to remember them.
- Type what you need to know onto a computer. Typing or rewriting notes is an effective way of memorising information for anyone who is a kinaesthetic learner.
- Work with a revision buddy. Each session you have together see who can think of the most ridiculous way of remembering something.
- Make up quizzes for you and your buddy to do.
- Drawing charts and diagrams can help you to memorise information, particularly if you are a visual learner.
- Use games like Maths Snap in which you have to match words and definitions and play a memory game, such as Grasping Grammar, in which you put the cards face down and have to turn over two cards at a time to see if they match.

REVISION BUDDIES

Get yourself a revision buddy! Revising alone can be boring, so team up with a revision buddy. Having a buddy can help you in a number of ways. You can test each other, compare notes on particular topics and practice explaining things to one another.

MAKE A GLOSSARY OF KEYWORDS

Explain what a glossary is and that making a glossary of keywords connected with a topic can help you to remember them. Encourage them to make a glossary of terms including any names of people and places of importance. You can build up the glossary as the study of a topic develops, asking individuals to write new words on post-its and to add them to a glossary on the notice board. Point out that glossaries are usually arranged alphabetically and discuss how the post-its should be arranged, so that the words appear in alphabetical order.

For example, when studying the Romans, they could include Julius Caesar, Pompeii, Hadrian's Wall, Jupiter and Mars along with words such as forum, villa, toga, amphitheatre, centurion and legion.

TEST YOURSELF QUIZZES

Explain that one way of helping yourself to remember key facts is to prepare a test yourself quiz to give to a revision buddy. Hold a competition by preparing a number of test yourself quizzes. Keep a record of your scores for each quiz.

MATHS SNAP

This is a game for two players. Invite the children to make two sets of maths cards by writing mathematical terms on one set of cards and their meanings on the other. They can then shuffle the two sets of cards together and play Maths Snap. Further examples of Maths Snap and other maths games can be found at www.active-maths. co.uk/games2/snap/index.html

GRASPING GRAMMAR

This is a game for two or more players. First, the players need to make two sets of 40 cards. Set A should consist of single words 10 of which are nouns, 10 verbs, 10 adjectives and 10 adverbs. Set B consists of 30 cards – 10 pronouns, 10 prepositions and 10 conjunctions.

Shuffle the cards and place them face down on a table. The players take it in turns to turn over two cards. If they turn over two cards that are similar, e.g. two nouns (Set A) or two prepositions (Set B) they pick up the two cards and have another turn. If they turn over two cards that do not match, they turn the cards face down again and the next player has a turn. When all the cards have been picked up, the winner is the person with the most cards.

PLANNING YOUR REVISION – SCIENCE

Show the children how to plan their science revision by using a self-assessment sheet to identify which topics they need to revise because they are less confident about those topics.

After they have identified which topics they need to concentrate on revising because they are less confident about them, the children need to list the topics in the order they are going to revise them and to draw up a timetable indicating when they plan to revise particular topics.

Science topics	Confidence ratings (5=very confident 1=not confident)				
Life processes and living things	5	4	3	2	1
Nutrition, diet and the digestive system					
Exercise, the heart and drugs					
Human growth and reproduction					
Plant life and growth					
Life cycles					
Classification of plants, animals and micro-organisms					
Food chains					
Evolution and inheritance					
Materials and their properties					
Characteristics of materials					
Rocks and soils					
Solids, liquids and gases					
Changing states					
Dissolving					
Reversible and irreversible changes					
Physical processes					
Forces and gravity					
Magnets					
Friction					
The Solar System (earth, sun and moon)					
Light and shadow					
Sound: how it travels and how it can be changed					
How humans see and hear					
Electrical circuits					
Electricity: conductors and insulators					

MNEMONICS

Explain what a mnemonic is – a word, phrase, sentence or verse that can help you to remember something that you need to learn.

Discuss how you can use these verses as a way of remembering significant dates.

To remember the date of the Discovery of America:

> In fourteen hundred and ninety two
> Columbus sailed the ocean blue.

To remember the dates of the Great Plague and the great Fire of London:

> In sixteen hundred and sixty five
> Hardly a man was left alive.
> In sixteen hundred and sixty six
> London burnt like a pile of sticks.

Discuss how sentences can be used as mnemonics.

Encourage the children to use mnemonics to help them remember key facts.

For example, the plant cycle can be memorised by using the following mnemonic:

Growing plants produce flowers that produce seeds that germinate into growing plants that produce flowers that produce seeds...

Similarly, to remember materials that are good conductors of electricity and ones that aren't they can use the mnemonic:

Monkeys conduct electricity but **P**igs, **S**nakes, **W**orms and **R**ats don't. This can help you to remember that metals conduct electricity but plastic, stone, wood and rubber don't.

A MATHS MNEMONIC

Put the word RUCKSACK on the board. Invite the children to suggest what the letters might stand for and build up the chart (below) explain that you can use this mnemonic to develop a step-by-step approach to a maths problem.

READ   Read the question.

UNDERSTAND   What does it ask you to do?

CALCULATE   What calculation is needed?

K   Can you estimate the answer?

SOLVE   Decide what strategy to use to solve the problem.

ANSWER   Have you shown how you worked out the answer?

CHECK   Check your calculations. Does the answer look...

K   correct?

CHAPTER 10
Helping children with special needs

Children with special needs require extra help and support to enable them to develop the study skills necessary for them to cope with their schoolwork. They often have difficulties because of poor memory, short attention span and low self-esteem. They may have difficulty grasping concepts and understanding what they have to do, be lacking in motivation and have problems organising their time. They may be easily distracted and find it hard to concentrate.

Each student is different, so how you intervene will depend on the type of difficulty they are experiencing and their particular learning style. Often, special needs children have problems fulfilling a task because they either cannot remember what they have been asked to do or because they do not fully understand the task. Such children can be helped by setting up a system of study buddies.

STUDY BUDDIES

Study buddies operate in pairs, which are carefully selected by the teacher, so that you pair more able children with children who have learning difficulties. When you are setting a task, before the children start you ask the pairs to take it in turns to explain exactly what the task requires them to do. This benefits even the most able children as well since it requires them to clarify what the task requires.

You can then ask the study buddies to write down what the task is or alternatively give the children with learning difficulties a written copy of what you want them to do. Children who are encountering problems with assignments and homework often do so because they cannot remember what they have been asked to do.

Also, many children, including those with moderate learning difficulties, can find it hard to grasp what they are required to do if the assignment is presented to them in general terms. Breaking down assignments into specific tasks will enable all children, whatever their abilities, to understand more clearly what they are being asked to do.

DEVELOPING ORGANISATIONAL SKILLS

Many children with special needs encounter difficulties in organising themselves and their time. It is important to remind children with learning difficulties to keep a homework diary and you need to check that they are doing so. They require help with basic organisational skills. The more organised they become, the more their schoolwork will improve.

They need help with organising their work, e.g. they need to be encouraged to keep the work they do on a topic in a coherent order, so that it can be accessed easily. If they get in the habit of putting the date on each piece of work, they can store the work chronologically. Alternatively, they can number the pages and store them in numerical order. If they are provided with large paper clips or ring binders they can store their work more easily than putting it loosely in a folder. Alternatively, special needs children can be shown how to store their work on a tablet.

Other organisational skills that will help them succeed include:

- Knowing what equipment they are required to take to school. Having a calendar on a wall in the kitchen or their bedroom on which they write what they need to take each day is a way of reminding them what to take.

- Putting such things as their calculator, pencils and crayons away in the same place, e.g. a drawer in the classroom or a shelf at home prevents time being wasted because they cannot find what they want.
- Always writing up on the board what you want the children to do acts as a reminder to those who have difficulty in following oral instructions.

DEVELOPING READING SKILLS

Children may have difficulty in comprehending what they read for a number of different reasons. They may be able to read a passage decoding all the words without understanding what the passage means. One reason they may not understand is that they are concentrating so hard to decode the words that they do not take in the meaning of the sentence.

Another reason for poor comprehension is that many children with learning difficulties are easily distracted and have short attention spans. They do not follow a paragraph of text consisting of a number of sentences, because their mind wanders and they lose focus.

Interactive apps can be useful with children with special needs who have short attention spans as a means of developing their concentration skills, e.g. Monster Hunt: The Memory Game in which the Monster hides in various places and the children have to remember where.

Children, who are experiencing difficulties with understanding what they read, can be helped by using a strategy that breaks down the information into units that can be absorbed one at a time.

Encourage the children to use the following strategy:

- Focus on one sentence at a time.
- Identify the subject of the sentence.
- Ask yourself: what does the sentence tell you about the subject?

- Are there any difficult words that you do not understand?
- Is it possible to work out what they mean from the rest of the sentence?

When starting a new topic, pre-teaching any new vocabulary that the children are likely to encounter is a useful strategy to help those with comprehension difficulties, particularly if you emphasise the words by using them repeatedly and posting them on a chart on the noticeboard.

Help the children to learn how to interpret sentences by looking for context clues both in a sentence itself and/or in the sentences that precede or follow it. Teach them to recognise the structure of a non-fiction text – the title, headings and sub-headings, which can give clues as to meanings.

DEVELOPING WRITING SKILLS

Children with learning difficulties may put off attempting writing tasks because they find them difficult and because of previous experiences of failure. They may have poor handwriting and often intervening by providing activities to improve their handwriting can lead to a marked improvement in the quality of their writing.

Organise a handwriting club. Invite children who dislike writing to join. Involve the children in setting up the club by encouraging them to make up rules for the club. They can give it a name and design a logo. They can make it exclusive by restricting membership. Adopt a multi-sensory approach, including games and activities such as are suggested by Handwriting without Tears www.hwtears.com/hwt

Children with special needs may be poor spellers and this may impact on the words they use by them restricting their choice of words to those they are confident of spelling directly. Such children can be helped to widen the vocabulary they use by providing them with lists of key words and phrases they could use when setting a writing task.

DEVELOPING NOTE-MAKING SKILLS

Children with moderate learning difficulties often have problems developing note-making skills. Giving them guidance by preparing note-making worksheets to use can help them enormously. Here are two examples:

Note-making paragraph by paragraph

What is the topic you are studying?

Write down two or more facts you learn from the first paragraph? Use your own words.

1.

2.

3.

4.

Write down any new words or terms you learned in the first paragraph.

1.   2.   3.   4.

Briefly say what the first paragraph was about.

Use the same approach to the other paragraphs in the text you are studying.

Guided note-making

This is an approach that helps those with moderate learning difficulties develop their note-making skills as it picks out the points they need to look for in texts about a topic.

For example, if they are studying an animal, you can use the guided notes sheet (below) to help them pick out the key facts from any text they read about the animal.

Guided note-making – Any animal

Animal name:

Appearance:

Countries where it is found:

Habitat where it lives:

Habits: what it eats:

when it sleeps:

which animals prey on it:

How does it reproduce?

Does it have any special features?

CHAPTER 11

Summary, assessment and other resources

Many children begin secondary school ill-equipped to deal with the demands that subject specialists make because they have not developed the study skills that are necessary. The activities presented throughout this book have been designed to remedy this situation and to facilitate the transition from primary to secondary school. The development of good study skills is an essential part of every child's education, enabling them to progress from primary school, through secondary school to further and higher education.

Children who have been taught how to plan investigations and projects, who know how to research information from texts and from the internet and to collect data through surveys and observations will be able to tackle problems effectively. The practice they have had in presenting their findings both orally and in writing will enable them to report their conclusions confidently. They will have learned too the value of talking in groups as a way of solving problems and learning what they think.

THINKING SKILLS

To sum up, they will have begun to develop the thinking skills that are embedded in the National Curriculum and that they will continue to develop as they progress through secondary school.

Enquiry skills: They will have learned how to identify problems and to ask relevant questions, to plan investigations, research topics and predict outcomes.

Information handling skills: They will have explored different ways of gathering information and of recording it and of collating and analysing data.

Reasoning skills: They will have learned the importance of giving reasons for their views and the conclusions that they draw.

Creative thinking skills: They will have begun to use their imaginations to think about how to solve problems by lateral thinking, suggesting and exploring solutions.

Evaluating skills: They will have developed the ability to evaluate information and to judge its reliability, gaining confidence in their assessments of its value.

ASSESSING STUDY SKILLS

The assessment sheets (overleaf) are designed to make the children aware of the skills that they have developed and to set themselves targets for the future.

The assessment sheet in this section can be used by both the teacher and the individual student. After putting a tick in the boxes to indicate what they think the individual can do, the teacher and the student can compare their assessments and discuss any differences.

STUDENT AND TEACHER STUDY SKILLS ASSESSMENT SHEET

Can explain what information s/he wants to find out about a subject and can frame what questions to ask.

Is able to plan an investigation and to draw up a schedule.

Is able to use appropriate resources to locate information such as the internet, encyclopedias, newspapers and magazines.

Is able to locate books in a library by subject, title and author.

Can find information in books using the contents and index.

Is able to use a dictionary to find the meaning of words and to check spellings.

Is able to use skimming and scanning in order to identify relevant information in a text.

Is able to identify key words in paragraphs and chapters.

Can paraphrase a passage and summarise key points.

Can make notes, numbering points and draw a mind map or flow chart as appropriate.

Can use apps to find information and make notes.

Is able to design and carry out surveys and questionnaires.

Is able to analyse data and draw conclusions.

Is able to report findings orally and in writing, e.g. by making a PowerPoint presentation..

Can reflect on the success of a project and identify targets for the development of study skills.

STRENGTHS AND WEAKNESSES IN SCIENCE

This self-assessment sheet can be used to identify a child's strengths and weaknesses in science and to set targets for improvement.

SCIENCE ASSESSMENT SHEET

Can create interesting questions to investigate on a variety of topics.

Can plan an appropriate enquiry, recognising and controlling variables.

Can make observations about what happens in an experiment.

Can take measurements using a range of scientific equipment.

Can record observations in a variety of ways, including drawing graphs, tables.

Can analyse test results, draw conclusions.

Can make predictions.

Can suggest further tests which are comparative and fair.

Can report and present findings both orally and in writing.

OTHER RESOURCES

Assessment sheets focusing on the skills required for giving presentational talks, which both teachers and students can use, are included in *Jumpstart! Talk for Learning* pages 27–28.

USEFUL WEBSITES

http://ngkids.co.uk This website is run by the National Geographic magazine and offers information about many subjects, including historical subjects, such as the Ancient Egyptians, as well as on the weather, climate change, the world's peoples and animals.

www.bbc.co.uk/schools BBC BiteSize, available at the BBC website offers introductory information on all subjects.

www.historylearningsite.co.uk is packed with information on history topics that is suitable for children in Key Stage 2 and Key Stage 3.

www.gyop.potato.org.uk This British Potato Council website has details of a project which teaches about growing potatoes, how they form part of a healthy diet and how they can be cooked. It also encourages the children to investigate how food was rationed during the Second World War, including a recipe for Woolton Pie, named after the Minister for Food, Lord Woolton.

www.practicalaction.org/schools has details of a number of challenges suitable for KS2, such as the plastics challenge which invites children aged 8–14 to investigate and suggest solutions to the problem of the amount of waste plastic produced globally.

www.topmarks.co.uk A useful source of information on all subjects.

www.nhm.ac.uk The website of the Natural History Museum.

http://rockwatch.org.uk The nationwide geology club for children provides information on all aspects of geology, runs competitions and organises field trips.

www.fiverchallenge.org.uk has details of an annual challenge supported by Virgin money which offers groups of children aged 5–11 a pledge of £5 per child to set up mini businesses to create products or services, which they can sell or deliver at a profit. The challenge encourages children to be creative, to learn enterprising skills and helps to develop their teamworking skills.

www.f1inschools.co.uk has details of an annual STEM challenge to research, design and manufacture the fastest race car possible made, for example out of card and powered in a particular way.

www.m-a.org.uk The website of the Mathematics Association has details of challenges aimed at the top 60% of children in Years 5 and 6 in England and Wales and P6 and P7 in Scotland, consisting of problems for them to solve.

www.sentinus.co.uk has details of STEM challenges suitable for Key Stage 2, such as to design and make a slow marble run, a lifting mechanism that can be fixed or stood on a table and that will grasp an egg and hold it above the table without dropping it. It includes details of the materials that you will need and that the children must use.

www.educationalappstore.com contains information on apps for all curriculum subjects.

www.apps4primaryschools.co.uk recommends apps for the primary classroom.

www.belb.org.uk has a booklet on infusing thinking skills and personal capabilities into numeracy teaching to Years 3 and 4.